Instant Notes on GENETICS, PLANT BREEDING & SEED TECHNOLOGY

THE AUTHORS

B. Hari Vara Prasad is working as faculty in Dr. Y. S. R. Horticulture University. He had also been awarded ICAR-JRF and SRF fellowship. He did his M.Sc (Ag) in Genetics and Plant Breeding from PDKV, Nagpur. He did his Ph.D (Ag) in Genetics and Plant Breeding from UAS, Dharwad.

P. G. Suresha is Ph.D scholar in Department of Genetics and plant breeding, UAS, Dharwad. He had also been awarded ICAR-JRF and SRF fellowship. He did his M.Sc. (Ag) in Genetics and plant breeding from UAS, Raichur.

Instant Notes on GENETICS, PLANT BREEDING & SEED TECHNOLOGY

– Authors –

B. Hari Vara Prasad

P. G. Suresha

2023

Associated Publishing Company®

A Division of

Astral International Pvt. Ltd.

New Delhi – 110 002

ISBN: 9789359191485

Published by : **Associated Publishing Company®**
A Division of
Astral International Pvt. Ltd.
– ISO 9001:2015 Certified Company –
4736/23, Ansari Road, Darya Ganj
New Delhi-110 002
Ph. 011-43549197, 23278134
E-mail: info@astralint.com
Website: www.astralint.com

Preface

Agricultural science is a broad multidisciplinary field of biology that encompasses the parts of exact, natural, economic and plant sciences that are used in the practice and understanding of agriculture. This book covers different important points which were asked in previous JRF, SRF and ARS-NET examinations.

In this occasion I heartly thank to all my batchmates who helped me to cover the bits of previous years. Lastly I take the blessing of my parents and grand parents for the success of his endeavour.

Authors

Important Points on Genetics and Plant Breeding

- Prebreeding : to utilize wild spp for expanding the genetic variability in primary gene pool to develop new cultivar with a broad genetic base
- FAD 2 genes in groundnut to increase high oleic and linoleic ratio
- Nagoya protocol-2010
- Globally -34 hot spots
- Megadiversity countries-17
- Bt cotton- EDV
- Sanger sequencing-600bp, Third generation sequencing-15kbp,
- Linkage mapping- recombination between marker and unknown genes
- Association mapping- correlation between marker allele and trait allele
- High R^2 value and less linkage disequilibrium- QTL between two markers can be used
- Transgenic variety in maize for drought tolerance-drought guard
- Hydroponics
- Pulling force in rice to study root strength
- MIR genes encode miRNA
- Agronaute proteins in RISC of miRNA
- miRNA are transacting elements
- Rht-B1b and Rht-D1b: dwarfing genes in wheat, gain in function, GA insensitive
- Rht8-GA sensitive dwarfing gene in wheat
- Sd1: dwarfing genes in rice
- Genetic slippage

- High heritability-response to selection
- Genetic assortative mating- inbred line development
- Genetic dissortative mating- interspecific hybridization
- Phenotypic assortative mating- recurrent selection- isolation of extreme phenotypes
- Phenotypic dissortative mating-disruptive mating- stable population
- Heterosis high in single crosses than double crosses
- Indica x japonica crosses high heterosis than other cross combinations
- Early generation selection-pedigree method
- Late generation selection-SSD, Bulk
- First cycle inbreds- single cross, open pollinated variety
- Source for salt tolerance in rice-pokkali rice
- Meuwissen et al. (2001) proposed genomic selection
- Finest fibre quality among cotton spp- G.barbadense
- Genes resistance to soyabean rust- Rpp series
- Less PCV and GCV indicates high uniformity in the population
- b(regression coefficient) values ranges from $-\infty$ to $+\infty$
- Non-significance of "b" indicates stability of variety over environments
- Significance of "b" indicates highly responsive to a particular environment
- Environmental index = mean of genotypes to a particular environment- over all mean
- Homeologous recombination occurs in groundnut but no multivalent formation
- GEAC- nodal agency in India in implementing EPA rules
- Bt egg plant-2013 in Bangladesh
- CBD-1992
- Cartagena protocol-2000
- Cartagena protocol- biosafety regarding transboundary movement of living modified organisms

- Nagoya protocol- administrative procedures and requirements that need to be taken in the event of damage by living modified organisms
- Nagoya protocol- japan
- Biotechnology regulatory authority of India bill-2012
- Cry1Ac gene- MON-531 event- American boll worm
- Registration of transgenic varieties under seed bill-2010
- National biodiversity Act-2002
- National biodiversity authority- Chennai
- Agriculture biosecurity bill-2013
- Plant genome savior community award-recognition to farmers for their role conservation of biodiversity
- Biomining- use of microorganisms to leach metals from their growth medium
- Biopiracy- a practice in which indigenous knowledge of nature, originating with indigenous people is used by others for profit without permission from and with little or no compensation or recognition to the indigenous people themselves.
- PVP-2001- Farmers are exempted for fee in registration of varieties
- High grain protein content- T.turgidum var dicoccoides
- Osmoregulation, antioxidant defense, phytohormones, high photosynthetic rate at low water potential- Drought tolerance
- High biomass before anthesis leads to high harvest index
- NBPGR-1977
- National gene bank-NBPGR
- Sunflower hybrid: BSH-1: CMS234 x RHA-274
- Bajra hybrid: HB-1: Tift 23A x J.14
- Ridley, Lerma rojo, sonorojo-64: direct introduction in wheat
- Agro ecological zones- 8
- Safflower, sunflower introduced oilseed crops in India
- Plant introduction reporter- published quarterly with information on introduced and indigenous collection
- Parbhani kranthi- pusa sawani x E.C
- Punjab padmini- pusa sawani x E.C

- Himani- Atlas-54 x NP-109
- Oryza nivara, oryza rufipogon most widely grown wild relative in India
- ICRISAT- 1973
- AICRP on under uitilized crops- 1982
- AICRP on Arid legumes-1997
- UPOV 1978- No farmer exemption
- UPOV 1991- Farmer exemption
- GATT- Geneva -1948
- Novelty, Distinctiveness, Uniformity, Stability- protection of plant varieties
- In *in vitro* conservation mannitol is used as osmotic agent
- Mannitol – increases shelf life of shoot tip culture by 12-14 months
- CGIAR centres-12
- Dehydration postponers-Drought escape
- Drought escape- rolling of leaves, covering bud tissues with scales
- Stomatal conductance-porometer
- Infrared thermometer-canopy temperature
- Watch tower genes- major genes signaling for disease resistance
- Pollen irradiation by UV rays
- Penetration –γ rays> x rays > β rays > α rays
- Dimethyl sulphoxide- enhances absorption of chemical mutagen
- Auerbach – discovered new chemically induced mutants in drosophila
- Sharbati Sonora and pusa lerma- mutants in wheat from Sonora – 64 and lerma rojo
- T-141 to jagganath variety in rice through mutation
- Balanced lethal system- Darlington
- Armadillo- First triticale variety (2R-2D substitution)
- Wright- Random genetic drift
- Co 205- first interspecific hybridization in sugarcane
- Lamda phage- 30Kbp

- Cosmid – 40 Kbp.
- BAC- 105-350 Kbp
- YAC- 400 to 700 Kbp
- LOD score – 2 (precision of linkage)
- FAD 3 loci- omega 3 desaturase gene
- Sr-2, Lr-34, Yr-18: durable resistance genes for rusts
- Copy and paste transposons: LINES and SINES
- Cut and paste : retroposons, Ac-Ds
- Article 15: CBD
- Chromosome 11: Xa^3, Xa^4, Xa^{21} genes in rice for bacterial blight resistance
- Long day and cool temperature for potato flowering
- Short day, cool weather and warm nights, high humidity for sugarcane flowering
- Ma_1 to Ma_4 maturity genes in sorghum
- dw_1 to dw_4 : dwarfness in sorghum
- Double dwarf in sorghum: $dw_1dw_1dw_2dw_2dw_4dw_4$
- dwarf sorghum: $dw_1dw_1Dw_2Dw_2dw_4dw_4$
- dw3 is unstable dwarfing gene in sorghum
- FAP – decrease palmitic acid, FAD- increase oleic acid: recessive mutations
- 'Ps' mutant in meiosis-I and 'os' mutant in meiosis-II of potato
- 10 Full sib matings = 3 selfed generations to achieve homozygosity
- $X_{AB(observed)} = gca_A + gca_B + sca_{AB,}$ $X_{AB(expected)} = gca_A + gca_B + mean$ and $X_{AB(observed)} - X_{AB(expected)} = sca_{AB}$
- $b_{op} = ½ h^2$
- $b_{op} = h^2$
- First transgenic- tobacco-1983
- Gametocidal chromosomes- Aegilops caudate
- Seeds act-2004 adopts UPOV-1978 of farmers exemption
- 15yrs-field crops, 18yrs- tree crops;
- 20yrs- field crops, 25yrs- tree crops maximum duration

- Arabidopsis genome project-2000
- Human genome project-2001
- Rice genome project-2002
- WTO is successor of GATT
- Thomas Roderick-Genomics
- $r^2 = 0$ indicates linkage equilibrium
- $r^2 = 1$ indicates linkage disequilibrium
- Markers less than 10Kbp from loci strongly associated with phenotype for all the traits
- $R^2 > 10\%$indicates more of common alleles
- Quarternary gene pool: Transgenic approach
- Genetic enhancement- prebreeding- Jones
- Generative apospory- archesporial cell
- Somatic apospory-nucellus, integument
- Land race- heterozygous and heterogenous population
- Convergent improvement: A(1:99) x B(99:1) - F_1 x A; F_1 x B
- Hull-over dominance term
- Genetic diversity maintained in self pollinated crops: multilines, varietal blends, mass selected varieties
- Diallel selective mating is common approach for creating variability in self pollinated crops.
- True breeding value- progeny test
- Doubled haploid line often inferior to inbred line –high level of homozygosity
- Sister alleles migrate to the same gamete- double reduction
- Quasi-diploid-2n + 1-1
- Polygenic traits have high breeding value- high additive effects
- Frequency dependent selection- alleles with low frequency are selected to maintain balanced polymorphism
- Phenotypic stability of a genotype- Adaptability
- Suitability of a variety over wide range of environments-stability

- Tissue culture techniques, GMO's are under ban for organic plant breeding
- Oryza breviligulata originated from oryza glaberrina
- Oryza sativa originated from oryza rufipogon
- Rate of fixation of unfavourable alleles at QTL's with small effects is higher under MAS than under phenotypic selection
- Multiplex ratio is low for RFLP
- CAPS derived from RAPD to RFLP : locus specific-co dominant,
- SSCP- co dominant
- SCAR – RAPD to RAPD
- Heteroduplex analysis- co dominant and locus specific
- Heteroduplex analysis- SNP identification
- Limited scope for back cross mapping population if the trait is dominant marker
- Rice and corn first transformed and regenerated by electroporation method
- Guha and maheswari- anther culture in datura
- STS-single copy in a genome
- Fixation of alleles in small population- Founder effect
- Fixation of rare genes- Founder effect
- Bulk segregant analysis-Michelmore et al.
- NIL's- fine mapping, tagging of traits and functional genomic studies
- Dwarf genes lead to premature stop codon in della proteins so that they remain as constitutive repressors.
- DELLA proteins bind to GA responsive genes which results short stature
- NIL's, RIL's and D.H's- 1:1 ratio for both dominant and co dominant marker
- Anti nutritional factors: Phytic acid obstructs ca and Fe absorption
- Anti nutritional factors: Saponins – steroidal glycosides high in pulses and oilseeds
- Anti nutritional factors: Cyanogenic glycosides- legumes

- Anti nutritional factors: Chlorogenic acid- safflower meal contains high levels
- In 1925 jones identified CMS source in onion
- In 1954 stephens and Holland identified CMS milo cource in sorghum
- In 1956 First sorghum hybrid- CSH-1: CK60A x IS 84
- In 1972 first sunflower hybrid
- A-lines and R-lines should be grown perpendicular to each other during hybrid seed production.
- Autosomal locus – 2 generations to achieve equilibrium
- Allosomal locus- equilibrium attained slowly because allelic frequencies are not same in the two sexes
- Inbreeding results in change of genotypic frequencies not in allelic frequencies
- Population size decreased from 4^n to 2^n in SSD and bulk breeding method
- In case of back cross method population size decreased from 2^n to 1.
- Genotypes decreased from 3^n to 2^n in SSD and bulk breeding method
- The harvest index of semidwarf cultivars is around 0.45 to 0.55
- Field resistance for a pathogen is because of horizontal resistance
- Seedling resistance for a pathogen is because of vertical resistance
- Indica x japonica eg: mahsuri varieties
- Low amylose- sticky grain Eg: Japonica
- High amylose- Hard grain
- Low amylose-high gelatinization temperature
- Shakti – high lysine and hard endosperm
- Rattan and protina – soft endosperm
- Teosinte : primitive maize (Zea perennis) x Tripsacum
- “ae” and “Wx” – 100% amylopectin
- PGMS- long day > 13.75 hrs –sterile
- PGMS- short day > 13.75 hrs- fertile
- Sporophytic apomixes-apospory, adventitive embryony
- Gametophytic apomixes- apogamy, parthenogenesis, diplospory

- Parthenogenesis and pseudogamy-non autonomous endosperm
- Black gram secrete malic acid, oxalic acid and citric acid
- Effect of pollen on expression of endosperm in maize is xenia
- Effect of pollen on maternal tissues (nucellus, integuments) of female parent is metaxenia
- Pusabold, varuna, kranthi are varieties of mustard
- If inbreeding depression is less in F_2 then it is said more of fixed additive x additive. Feasible for early generation selection.
- Thumb nail pressure: high oil content types pressed easily
- Low micronaire value- very fine fibre
- Halo length- length of fibre with attached seed
- Mean fibre length- mean of the length of all the fibres
- Fibre strength- g/tex (tenacity)
- In F_2 3 (susceptible):1(resistant) ratio indicates disease resistance is governed by recessive gene.
- First commercial multiline variety was Miramar 63 (stripe and stem rust resistance)
- Super race- is a pathotype possessing virulence against a large number of useful host resistance genes in combination.
- KSML-3
- Sonalika multiline 1- (6)
- Multiline kalyan sonalika 11- (8)
- Multiline – clean crop approach and dirty crop approach by Marshall
- Gene pyramiding- vertical resistance genes
- Sr-26: Agropyron elongatum – durable resistance
- Yr-8: Aegilopsis comosa
- Lr-9 & Lr-24: Aegilopsis umbellulata
- Lr-9 + Lr-24 – no virulence
- Sr 2 + Sr 36 – durable resistance
- Lr-28: Aegilopsis elongatum
- Lr-26: rye

- Gene deployement-strategic release of varieties with vertical resistance gene in the puccinia path of rust pathogen
- Yellow rust survives only in north India because spores requies less than 16 °C temperature for germination
- Biffen –genetics of disease resistance on yellow rust-controlled by monogenic recessive gene
- Hypersensitivity – host cells die in response to pathogen attack so that progress of pathogen is restricted.
- Immune – complete resistance/freedom from disease symptoms
- Avirulence – failure of pathogen to attack and produce disease symptoms on the host
- Ontogenic resistance- plants will react either resistance or susceptible based on its age
- Ability of a variety to differentiate between races on the basis of disease reaction- differential varieties.
- Ability of pathogen to infect and cause symptoms of varying degree- Aggressiveness
- Multilines – synthetic horizontal resistance
- Recessive epistasis- supplementary epistasis-9:3:4
- Double recessive epistasis-9:7
- Dominant epistasis- masking gene action-12:3:1
- Double dominant- duplicate epistais-15:1
- Turner syndrome-monosomy of X chromosome
- Down's syndrome-trisomy of chromosome 21
- Patau's syndrome- trisomy of chromosome 13
- Edward's syndrome- trisomy of chromosome 18
- Kliefelter's syndrome-XXY
- Chromosome 4: pod shape, plant height, position of inflorescence
- Chromosome 7: seed shape
- Chromosome 1: cotyledon color
- Double steel and tera petkus- autotetraploid crop varieties of rye
- Vermillion eye- unreduced female gametes- chromosomal theory evidence

- Chaperonins-prevent denaturation and prevent incorrect folding of proteins during stress conditions
- Uniform expressivity- all individual which carry gene show same expression
- Variable expressivity- all individual which carry gene do not show same expression
- Conjugation- Lederberg and tatum
- UUU-phenyl alanine
- Khorana- alanyl transfer RNA
- Nirenberg – genetic code
- Histone proteins are devoid of tryptophan
- H_1- lysine rich
- H_{2a} and H_{2b}- slightly lysine rich
- H_3 and H_4 – arginine rich
- Chiasma terminalization- diplotene stage
- Constitutive heterochromatin- centromere
- Moderately repetitive- 2 to $>10^5$ copies
- Moderately repetitive- transposable elements, rRNA, Histones, ribosomal proteins
- Highly repetitive- $>10^5$ copies
- Highly repetitive- centromere & telomere
- In male drosophila no crossing over- no synaptonemal complex
- Philaldelphia 22 – deletions in chromosome
- Cry of cat syndrome - deletions in chromosome 5
- Anaphase – shortest stage in duration
- S-phase in interphase- DNA duplication
- In prokaryotes 70S ribosomes divided into-50S and 30S
- 50S- 23S and 5S; 30S- 16S RNA;
- In eukaryotes 80S ribosomes divided into-60S and 40S
- 60S-28S, 5.8S and 5S; 40S-18S
- Adjacent-II segregation: homologous centromeres move to same pole

- Adjacent-I segregation: non-homologous centromeres move to same pole
- Alternate segregation: translocated and normal chromosome move to same pole
- Maize pollen grain with waxy allele show brown on staining
- Maize pollen grain with non waxy allele show deep blue on staining
- Genes within the inverted portion are linked because of inversions are cross over suppressors
- About 75% repetitive DNA in human genome is of LINES and SINES
- Comparative analysis of genome sequences of related spp. to detect orthologous DNA sequences is known as Phylogenetic footprinting
- When the gene order in a given region of the genome of two organisms is comparable-synteny
- BLAST- database to assign new gene or protein
- Full complement of RNA molecules produced by the genome-Transcriptome
- Genome annotation-process that identifies genes, their regulatory sequences and their function
- Open reading frame- coding region of gene
- Orthologous genes- homologous genes in different organisms
- Orthologous genes arise due to vertical descent during evolution
- Paralagous genes- homologous genes with in organisms but gene products with non identical function
- Paralagous genes arise due to duplication during evolution
- Differences and similarities in genome structure and organization of different organisms- comparative genomics
- Storage and analysis of genome information through a database-Bioinformatics
- Crossing over at introns leads to different combination of exons i.e new domain proteins- exon shuffling
- Genetic exchange between different evolutionary lineages-horizontal gene transfer
- The genome has gene rich regions separated by gene poor regions often called gene deserts

- Sequencing of genomes a)clone by clone b)short gun method
- DNA fragmented into BAC clones of 80-100Kbp and then each BAC clone fragmented into cosmid clones.
- Expressed sequence Tags- can be used as probes for isolation of concerned genes from the genomic DNA.
- 97% genome similarity between human and mice
- M.germitalium-0.58Mbp
- E.coli- 4.64Mbp
- H.influenzae- 1.83Mbp
- Yeast- 12.8Mbp
- Round worm- 97Mbp
- Arabidopsis-130Mbp
- Drosophila-180Mbp
- Rice-415Mbp
- Human -3300Mbp
- Sorghum- 750Mbp
- Maize -3000Mbp
- Barley -5000Mbp
- Wheat -16000Mbp
- Amount of DNA in pg with in a haploid nucleus- C-value
- C –value paradox- amount of DNA need not to be very with increase in complexity of organism.
- Haploid -1C
- Diploid -2C
- On average one gene per 100-150Kbp but varies with size of organism.
- Multigene families- groups of genes of identical or similar sequence
- Suppressor mutations- one of the termination codons being recognized as a codon for an aminoacid
- Genome size varies between 10 to 10^5Mbp
- Pseudogene- integration of cDNA produced from mRNA of functional gets integrated into same or different chromosome.

- Mitochondrial genome free of introns
- Genome wide repeats are appears to be derived from transposable element.
- Hershey and chase- DNA as genetic material in T_2 phage of E.coli
- Frankel, conrat and singer- RNA as genetic material in TMV
- Genes are located in GC rich regions
- Evidence of semiconservative DNA replication- $cscl_2$ density gradient centrifugation
- Origin of replication sites contain AT rich region.
- Phosphodiester bonds in purines 9th carbon of nitrogen base and 1st carbon of sugars
- Phosphodiester bonds in pyramidines 1st carbon of nitrogen base and 1st carbon of sugars
- DsRNA is in A-form
- DsDNA is in B-form
- E.coli chromosomes were shown to exist as theta shaped structure during replication.
- Replication is bidirectional confirmed by studies on T_7 E.coli phage
- Mammalian mitochondria- displacement loop DNA replication
- DNA pol-I: digestion of RNA primer and filling of gaps generated in new strands
- DNA pol-III: DNA replication invivo
- Klenow fragment of DNA pol-I invitro replication
- DNA pol-I: 5^1-3^1 exonuclease activity- proof reading to reduce errors in replication
- Rolling circle replication: Single stranded bacteriophage ØX-174
- Pilot proteins- these proteins decide whether viral DNA to under go replication or transcription.
- Deamination of 5methyl cytosine = thymine
- Linkage mapping- 2 point/3point test crosses
- Cytogenetic mapping- FISH and GISH
- Molecular mapping-DNA markers

- First eukaryotic organisms sequenced – yeast, Arabidopsis and drosophila
- Gene cluster: members of gene family are together at one location.
- Illegitimate recombination-when the two non homologous DNA segements contain the same Insertion Sequence element recombination can occur between their IS elements.
- Dissociator- it activates only in presence of Ac element which means transposition.
- Locations in the map of markers like SSR are called sequence tagged sites
- Linkage of marker is detected by statistical test developed by morton- Lod score method
- Lod score = $\log_{10} \frac{\text{Probability of linkage}}{\text{Probability of independent assortment}}$
- Lod = 3 (10^3) 1000 times probability of linkage
- Many DNA markers are clustered near centromere region because crossing over is less near centromeric region (heterochromatin region)
- All the known nuclear introns of eukaryotes begins with dinucleotide GT and end with the dinucleotide AG.
- Histone mRNA's do not have poly(A) tail
- The base sequence of mRNA is changed after transcription- RNA editing
- 5^1 leader sequence : shine dalgarno sequence (hexamer) is ribosome binding site i.e pairing between mRNA and rRNA in ribosome.
- UAA-ochre
- UAG-Amber
- UGA-umber/opal
- Wobble hypothesis – 3rd bp of codon of mRNA in 5^1-3^1 and 1st bp of tRNA in 5^1-3^1 direction.
- UGA act as codon for tryptophan in mitochondria
- Polypeptide chain forms in N-C direction
- Effectors are of two types inducers and co repressors in gene regulation

- LAC operon-inducible negative
- Tryptophan operon- repressible positive
- When inducer makes repressor protein inactive on binding to it-inducible positive regulation.
- Co repressor binds to repressor and makes it active to stop transcription
- Labeling of nucleic acids are used chiefly to produce radioactively labelled probes-Nick translation
- Probes in DNA fingerprinting-Mini/microsatellites
- Alu I- AG CT –blunt ends
- Nla III- cleavage outside the sequence
- Emil fisher- peptide linkage
- Annealing temperature depends on the length and the GC content of primer
- Anti sense DNA strand = sense RNA
- RNA pol I- rRNA
- RNA pol II- mRNA
- RNA pol III- tRNA
- Sigma factor in RNA pol involves in stable binding of holoenzyme to promoter
- Sickle cell anaemia- glutamic acid replaced by valine at 6th amino acid position
- Co linearity of polypeptide sequences and mRNA sequences- "Yanofsky"
- Allelic frequency of recessive allele after migration $q_1 = q_0 + m(q_m - q_0)$
- At mutational equilibrium $P\frac{V}{U+V}; q\ \frac{V}{U+V}$
- The proportion of a population that is lost due to selection is called genetic load
- Genetic death- it refers to the inability of individual to reproduce due to sterility
- The fixation of genetic variants in a population through such molecular process is called molecular drive

- Canalization – development of character is unaffected by genetic variations and environment.
- Complete linkage- appearance of parental types when genes are located on same chromosome.
- Variation in seed color in wheat and oats was not continuous. Such polygenes showing discrete variation are called quasi-quantitative characters.
- $\chi 2$ –cal >0.05 probability level-null hypothesis accepted.
- At the beginning of diplotene synaptonemal complex disappears and chiasmata becomes visible
- Intrinsic terminator due to formation of hairpin loop- Attenuation
- In genetic imprinting demethylation must occur in embryo while methylation has to take place during gamete formation.
- Paramutation refers to a change in the function of an allele of structural gene (paramutable gene)
- Paramutable allele undergoes methylation under the influence of paramutagenic allele
- Linkage decay = f(Ab x aB) – f(AB x ab)
- Genetic homeostasis and canalization – conceal and preserve genetic variability
- Harlan and dewet gave concept of gene pool
- According to Harlan 3 major centres and 3 non-centres
- Passport data- information at the time of germplasm collection
- TN-1 – first high yielding semi dwarf variety with sd-1 gene from dee-geo-woo-gen
- IR-8 (peta x dee-geo-woo-gen)
- French bean- Kentucky wonder
- Okra-pusa sawani
- Cow pea- pusa barsati
- Cabbage- pusa drum head
- Brinjal- pusa purple long, purple round
- Protandry-carrot and celery
- Often cross pollinated crops- cotton, pigeon pea, okra, sorghum

- Negative inference- one cross over independent of other cross over
- Egyptian clover-berseem
- Jaya – TN-1 x T-141
- CMS source in rice – O.sativa f. spontanea
- IR-8 and jaya –primary introduction for rice in India
- CIMMYT-semidwarf varieties of spring wheat - introduction in India
- USA- semidwarf varieties of winter wheat- introduction in India
- % homozygosity AA- $(2^m-1)^n/(2^m)^n$
- The average recovery of recurrent parent = probability of eliminating undesirable allele from donor parent (%) = $1-(1-p)^{m+1}$x 100
- Frequency of genes from donor parent in any back cross generation is $(½)^{m+1}$ x 100
- High inbreeding depression in maize
- Low inbreeding depression in onion, sunflower, rye, pearl millet
- Maximum heterosis r_a=0 ((complete dispersion of alleles); (d) and (dd)-same sign
- Minimum heterosis r_a=1 (complete association of alleles); (d) and (dd)-opposite sign
- F= VB_2-VB_1; 'F' is positive P_1 parent has more dominant alleles
- Degree of dominance $\sqrt{2VD/VA}$
- For K alleles at loci then No of genotypes = k(k+1)/2, No of homozygotes = k; No of heterozygotes = k(k-1)/2
- For 'n' loci total No of genotypes = $(k(k+1)/2)^n$
- Free variability- AABB
- Homozygote potential variability – AAbb x aaBB
- Heterozygote potential variability – AaBb
- Genotypic composition in any selfed generation 1 + $(2^m-1)^n$
- The differential effect of gene in different genetic backgrounds can be studied by developing isogenic lines in different genotypes.
- Opaque 2 mutant in maize increases lysine content in kernel but cause late maturity because of pleiotrophy

- A random mating population segregating for 'n' loci is capable of producing 2^n inbred lines
- $N = 2^n - n - 1$: possible synthetics
- Synthetics are expected to have selfed progenies and the ratio of selfed to crossed = 1/n-1
- Poly cross test – tysdall
- Vybrid – cross between two facultative apomicts
- Diplontic selection – unmutated cells over grow than mutated cells
- Primex –X-ray mutant mustard variety
- Canola – double zero rape seed
- Cytokinesis followed by first meiotic division gives unreduced gametes
- Colcemid –synthetic analog of colchicine
- Meristem explants – 0.2-0.4mm
- Antixenosis – true form of insect resistance
- Antixenosis- maximum selection pressure, bad host
- Plasmid – 10-15Kbp, cosmid- 40Kbp; Phage- 22Kbp; BAC-300Kbp; YAC-1000Kbp
- Vent polymerase- Thermococcus litoralis
- Recombination frequencies are converted into cM distance through mapping function called Haldane.
- $P = \frac{1}{2}(1-e^{-\mu})$ μ = frequency of chiasmata; p= recombination frequency
- One chiasmata between two genes produces 50% recombinants and the length of chromosome which on average has a value of 50 cM.
- Zeng – composite interval mapping
- NIL's – substitution mapping
- Type III error : significant association of QTL with a wrong marker is even more serious.
- $S.E_m = \sqrt{\frac{EMS}{r}}$
- $Syn_2 = \frac{syn1 - (syn1 - syn0)}{n}$
- $Syn_0 = \frac{A+B+C+D}{4}$; $Syn1 = \frac{AB + BC + CD + AC + AD + BD}{6}$

- In synthetics non-additive effects were exploited after accumulation of favorable additive effects.
- Lox sequence : blocking sequences in terminator gene technology
- In India four classes of seed under seed multiplication chain i.e breeder, foundation, registered and certified seed
- Lattice design – incomplete block design with perfect square of whole numbers.
- LSD>CRD>RBD precision for error
- Augmented block design – minimum error degree of freedom- 12
- The maximum number of generation for seed multiplication is limited to 4; no generation limit for breeder and foundation seed
- Tag size – 12cm x 6cm
- Seed act- 1966 which is simple and farmers friendly
- Registration is not compulsory under seed act 1966.
- Seed control order- 1983
- Good cooking quality of oil (unsaturated): high oleic acid, high linoleic acid and low linolenic acid
- High stearic acid (saturated) - industrial purpose
- Waxy mutants – 100% amylopectin
- Albumins and globulins – major storage protein in oilseeds
- Maize – major protein fraction is Zein
- Barley - major protein fraction is hordein
- Pulses are deficient in methionine, cystine and tryptophan
- Cereals have high prolamine protein fraction which is deficient in lysine
- Additive effect – differences between homozygotes at a locus
- Centegener method – earliest method of single plant selection
- Anti nutritional factor : anti pyridoxine factor- linseed meal
- Goitrogens – soyabean and French bean
- Cyanogens – lima beans and chick pea
- Lectins – pulses and beans
- Trypsin inhibitors – beans, peas, chick pea

- Amioca – 0 % amylase
- High lysine mutants in cereals-reduced prolamine protein fraction
- Oleic acid- precursor of many unsaturated fatty acids
- Safflower – first crop in which genetics of individual fatty acids was reported
- An alternate approach to combine high yield with high lysine content by enhancing expression of chymotrypsin inhibitors.
- Low protein wheat contain high proportion of albumins and globulins which are comparatively high lysine content.
- High molecular weight- glutenins
- To improve sulphur containing amino acid globulins has to be increased
- Permanent mapping population – RIL's and D.H's
- Pusa giant napier – perennial napier x annual bajra
- Oleic acid, linoleic acid and erucic acid are major fatty cids in oilseeds
- XYZ system in wheat. X:2n+2R (restorer parent); Y:2n+1R (maintainer parent); Z:2n (male sterile parent)
- 2 Amino purine pairs with adenine; 5 Bromo uracil pairs with thymine
- Deamination of cytosine = uracils
- Deamination of adenine = hypoxanthine pairs with cytosine
- Replication slippage- same sequence copied twice or missed out
- Translational slippage- mRNA contains two or more copies of gene
- mRNA degradation – Rnase E and Rnase III- endonucleases; Rnase II- exonucleases
- maximum distance between two markers is 10cM
- first QTL identification was made by 'Sax'
- Genetically engineered male sterility-mariani (rapeseed and mustard)
- Annealing temperature is less by 1-2°c to melting temperature
- DNA microarrays- comparison of RNA content of different organisms
- DNA chips- screening of polymorphisms for SNP's

- Hind-II – first restriction enzyme
- RE-I – cleaves 1000bp away
- RE-III cleaves 20bp away on one side of sequence- asymmetric cuts
- Isochizomers – Hpa II and Msp I – methylated and unmethylated
- 4^n – frequency of recognition sites Eg: 4^2 any 2bp sequence occurs at every 16bp
- Vector should be ideally less than 10Kbp in size
- Most widely used vectors for gene cloning – plasmids
- Derivative of BAC- Fosmid vector
- Hybrid sorting- reverse breeding
- Pusa rice hybrid 10 : first released hybrid in rice
- Shaktiman – 1,2,3,4 are QPM hybrids
- The first CMS based pigeon pea hybrid GTH-1
- Opaque-2 composites- shakti, protina and rattan
- Drought avoidance – water savers and water spenders
- 1B long arm/1R short arm translocation in veery series of varieties of rye.
- The short arm of 1R carries genes for leaf rust (Lr 26, Sr 31, Yr 9)
- Winter hardiness- rye>wheat>barley>oats
- Non preference is due to Allomones/kairomones
- Rice stem borer – high silica content and lignified stem- Non preference
- Brown plant hopper – low aspargine- Non preference
- Maize stem borer – high aspartic acid - Non preference
- Pea aphid – low sugar content , low amino acid - Non preference
- Cucurbits – low cucurbitacins - Non preference
- Brassica cabbage aphid – low sinirgin - Non preference
- Alfalfa – high saponin - Non preference
- Sugar beet- low free sugars - Non preference
- Corn ear worm – toughness of husk - Non preference
- Cotton boll worm – necatrliness - Non preference

- Potato- sticky trichome exudates - Non preference
- Ecological resistance = apparent resistance = pseudo resistance = disease escape
- b=1, S^2d = 0 and high mean – Eberhart and Russell
- b=0 absolute stability; b=1 average stability
- Grid /stratified mass selection – control of environmental variation
- No of interactions $= \frac{mn!}{m!n!}$ 'm' genotypes and 'n' environments
- Cajanus cajanifolius ancestor of pigeon pea
- Cajanus scarabaeoides – CMS source in pigeon pea
- Pusa ageti variety of pigeon pea
- Soyabean occupies first in oilseeds in world and 3rd in India after groundnut and rapeseed
- In male sterile plants there is elevated levels of arginine and asparagines instead of proline in anthers.
- Sunflower – PET-1- cytoplasmic source- H.petiolaris
- In wheat photoperiod sensitive cytoplasmic genetic male sterility in Norin 26 which carries cytoplasm of Aegilops crassa
- NES pistillate lines : $>35^0$ c- pistillate condition
- Male gametocide 1 – Zinc methyl arsenate
- Ppd 1 and Ppd 2 - photoperiod insensitive genes in wheat
- Pusa basmati -1 : pusa 150 x karnal local – first high yielding photoinsensitive hybrid variety of basmati rice
- NP-4, NP-6, NP-12 : purline selection varieties of wheat
- Pigeon pea – 2011 – genome sequenced
- Suguna , H_4 and H_6 - G.hirsutum hybrids
- Savitri , jayalakshmi and varalakshmi - G.hirsutum x G.barbadense hybrids
- MAB – improved pusa basmati-1, improved sambha mahsuri-1, swarna sub-1 and vivek QPM-9
- When dominance is complete – RSGCA, RSSCA and RRS
- Kosambi mapping function assumes interference

- Haldane mapping function – do not assumes interference
- Improved pusa basmati –1 contains Xa 13 and Xa 21 genes for bacterial blight resistance
- Improved pusa rice hybrid –10 contains contains Xa 13 and Xa 21 genes for bacterial blight resistance and pi54 + piz 5 blast resistance
- Rice blast gene pyramiding – Pi 1(t), Pi 2(t), Pi 4(t), Piz 5 and Pita
- Sparrow variety in wheat showed resistance to all rusts
- Tightly linked marker should be less than 5cM
- Jenkins- RSGCA
- Hull – RSSCA
- Allomones – adaptive advantage to plants
- Kairomones - adaptive advantage to insects
- Fine structure of gene in r^{II} locus of T_4 even bacteriophage- Benzer-using complementation test
- Coconada-1 and Coconada-2 coloured cotton varieties in G.arboreum
- $C.V = \frac{S.D}{Mean} \times 100$
- Glyphosate herbicides prevents synthesis of glutamic and aspartic aminoacid
- Sulfonyl herbicides prevents synthesis of leucine, isoleucine and valine
- Roundup ready contains EPSPS enzyme which synthesizes glutamic and aspartic aminoacid
- Acetolactate synthase enzyme synthesizes leucine, isoleucine and valine
- Diplospory is due to unreduced embryosac by mitotic division or restituted meiotic megaspore mother cell
- One line system- apomixis
- Two line system – PGMS/TGMS
- Three line system – CGMS
- Autopolyploid hybrids have stronger heterosis than diploid hybrids because of high allelic diversity and high heterozygosity

- Autopolyploid inbreds show stronger inbreeding depression than diploid inbreds
- Biofortification – it is the process of increasing the bioavailable concentrations of an element in edible portions of crop plants through traditional breeding practices or modern biotechnology.
- Major micronutrient deficiencies – Vit-A, Fe, Iodine and Zn
- Dietary diversification, supplementation and biofortification- strategies to address micronutrient deficiencies
- Cry 1 Ac gene is under the control of CaMv 35S promoter
- Npt-II gene shows resistance to antibiotic kanamycin
- Bt hybrids contains Bt gene in hemizygous condition
- ELISA test- to detect the presence of cry 1 Ac protein
- PCR amplification using primers and southern blot analysis using probes helps to distinguish Bt and non Bt
- Crystal proteins dissolve in alkaline medium and bind to receptors in gut which the condition is absent in eukaryotes
- G.hirsutum – long staple cotton
- G. barbadense - extra long staple cotton
- TSP-10R exotic pistillate line in castor
- VP-1 indigenous pistillate line in castor
- GMS source in cotton : Hisar source and Akola source
- AAH-1 : first GMS based hybrid in India in diploid cotton
- Greg male sterility – duplicate recessive genes ms_5 ms_5 ms_6 ms_6
- Diploid interspecific hybrids – DDH-2, DH-7, DH-9
- Intra hirsutum hybrids – Bunny, dhanalaxmi, savitha and H_4
- D_2 CMS – G.harknesii
- Doak's method of emasculation and pollination – cotton
- B73 x Mo 17 heterotic population in maize in world
- Suigeneris system – a system of its own kind
- Technologies which prevent germination of seeds of produce are not permitted in India- Gene Use Restriction Technologies (Terminator and traitor gene technology)

- GM variety/hybrid – Essentially Derived Variety
- ICPH-8 : MS prabhat DT x ICPL 161 in pigeon pea
- Morphological characterization of male sterility 1 gene in pigeon pea is translucent anthers and ms 2 gene is arrow shaped non dehiscent anthers
- GMS lines with male sterile 1 gene in pigeon pea – MS 3A, MS 4A and MS prabhat DT
- CMS first discovered in Brassica napus
- Brassica rapa – cross pollinated due to sporophytic self incompatibility
- Brassica napus and Brassica juncea highly self pollinated
- Ogura CMS source is commonly and widely used in Brassica
- Barnase – bacterial RNase with TA29, tapetum specific promoter and PPT acetyl transferase gene which shows resistance to Phosphinothricin herbicide.
- The insertion of transgene at allelic position was achieved by using a site specific recombinase in to pre existing lox-site.
- Safflower – premium oil quality rich in poly unsaturated fatty acid with > 78 % linoleic acid
- DSH-129 : India's first GMS based safflower hybrid
- Stable cytoplasmic source in bajra is A_5
- Orf-H522 sequence associated with male sterility in PET-1 cytoplasm of sunflower
- GCH-3 hybrid in castor: TSP 10R x JI 15
- VP-1 : most versatile pistillate line being used in hybrid breeding programme and also development of new pistillate lines.
- Sex reversal variants is S-type in castor
- N-type is controlled by recessive gene for femaleness
- Over dominance – interallelic interaction and repulsion phase linkage
- F = 0 : open pollinated, F = 1 : self pollinated
- The homozygosity of individual depends on allelic similarity in the gametic constitution during mating of both parents – identical by descent

- Methylation of DNA is gradually accumulated while selfing of crop spp which is afterwards released/repatterned when selfed genotypes are mated to develop hybrids.
- Indica x japonica > indica x javanica > japonica x javanica – order of heterosis in rice
- Petalloid male sterility in cotton G.arboreum
- Barley CMS source – H. jubatum
- Self sterility may lead to production of flowers for longer period which is advantage for ornamental spp
- 4 QTL's in sorghum for stay green nature – Stg B, Stg 1, Stg 3 and Stg 4.
- Chromosome segment substitution lines – to explore potential of wild spp chromosome segment were transferred into elite genetic background to evaluate several agronomic traits by using QTL's
- RSL's-Recombinant segregants lines (BC_4F_2, BC_5F_2, BC_6F_2)
- MAGIC – Multi parent Advanced Generation Inter cross population
- Indica MAGIC- intercrossing of 8 indica founder lines
- MAGIC global - intercrossing of Indica MAGIC and japonica MAGIC
- Fragrance in basmati rice is due to 2 acetyl 1 pyrrolidine
- Floury – soft endosperm, flint – (very hard) and dent (semi hard) in maize
- Riso 1508- high lysine mutant in barley
- Negative pleiotropic effects in mutants are broken by recurrent selection method
- Glutinous – amylopectin
- High amylose – industrial application
- Pasta is prepared from Triticum turgidum
- Porridge – low amylose and high amylopectin
- SDS PAGE- separation of proteins through vertical gel electrophoresis
- Rainfed – drought tolerance and recovery mechanisms
- Dryland – drought escape and drought avoidance
- Post fertilization barriers high in indica x japonica cross but heterosis is also high

- Commonly utilized cytoplasm source in sorghum is milo (A_1)
- Sorghum progenitor – Sorghum arundinuceuin
- Broom corn millet – most efficient drought tolerant
- Wild abortive cytoplasm – O. sativa f. spontanea (japonica)
- Stephens and Holland developed CMS line when they made cross combination of milo x kafir.
- RIL's contain high V_A with negligible V_D. So selection is more effective than F_2/ F_3 (.V_A + V_D + E)
- Solanum demissum – source for late blight resistance in potato
- Phureja technique in potato is to produce haploids
- Closest relative of maize – Tripsacum
- Endosperm balance number hypothesis explains triploid block in potato
- Sugary enhancer (se), shrunken -2 (Sh 2)- super sweert corn
- B chromosome – reduced fertility, decreased vigor, aborted pollen, defective kernels.
- Hard wheat – bread making, soft wheat – cookies
- Double haploid technique – development of inbred lines
- OPV x OPV do not produce heterotic hybrid because of less allelic difference.
- Maximum heterosis can be exploited between two inbred lines because of high allelic difference of 1 and mean allelic frequency of 0.5
- Endoploidy – cell gets differentiated after 'S' phase without entering mitosis phase and may continue replication.
- Dedifferentiation – differentiated cells becomes meristemic
- Proembryo-globular-heart-torpedo-cotyledonary stages of embryo development
- Appearance of similar alleles in one gamete because of chromatid segregation
- Allogenotypic competition- competition with in a segregating population in which different genotypes occur.
- Fatuoides – spontaneous mutants in oat cultivars that resemble wild oats

- Cross resistance – resistance associated with a change in one genetic factor that results in resistance to differential chemical pesticides that were never applied.
- Disomic polyploid – regular bivalent pairing occurs only between strictly homologous chromosomes
- Pistilloidy – development of flowers bearing pistils but no stamens
- Position effect – phenotypic modification resulting from the arrangement of genes relative to each other.
- Pseudogamy – apomictic seed set induced by pollination with out fertilization.
- Genetic drift – shift in gene frequencies
- Nick – synchronization of the receptivity of female organ to the maximum pollen load of the pollinator for cross fertilization
- Semigamy – abnormal fertilization in which the male gamete fertilizes the egg but not fuse with egg nucleus in cotton.
- Gentic stock – genotype maintained as a preferred parent in breeding programme but which may not be released as cultivar.
- Zertation – restricted or reduced growth speed of pollen tubes through genetically different stylar tissues.
- Prepotency – individual to produce identical progeny and also to individual
- Genetic assortative mating increases prepotency
- Breeding value refers to the value of genes transferred from a parent to its progeny
- Genetic value refers to the value of genes to the individual itself.
- If mean value of F_2 greater than F_1 it indicates lack of dominance
- Maize flower – 3 anthers
- Cleistogamy – wheat, oat, barley, soyabean
- Wild barley – Hardeum bulbosum
- Nicotiana tabaccum (SSTT) - Nicotiana sylvestris (SS) x Nicotiana tomentosiformis (TT)
- Triticum monococcum –einkorn wheat, Triticum diococcum – emmer wheat Triticum turgidum –durum wheat

- Sears developed cytogenetic aneuploid stocks from standard cultivar Chinese spring wheat
- Jenssen – diallel selective mating system
- Maize , sugar cane– short day plant
- Gene pool – I : cultivars
- Gene pool – II : wild progenitors
- Gene pool – III : Alien or other spp
- Balanced tertiary trisomic in barley on selfing gives 70 % sterile diploids and 30 % trisomics
- Xa13 and Xa 21 genes were introgressed from IRBB 55 into genetic background of pusa basmati 1
- INDUS – extant notified varieties database
- Nitrogen use efficiency : nitrate transporters play a role in accumulation of nitrogen in plant
- CcRpp 1 : gene from pigeon pea confers resistance to soyabean rust (Phakospora pachyrhizi)
- CcRpp 1 belongs to Nucleotide Binding Site-Leucine Rich Repeats
- Waxy corn (more of amylopectin)- recessive mutant which silences endosperm granule bound starch synthase enzyme which makes amylase.
- 'Sub -1' QTL on chromosome 9 explains 70 % of variation of submergence tolerance in rice
- "CLAVATA 3" – giant fruit size in tomato because of increase in stem cells
- R-Avr interaction gives host resistant response
- Rice blast gene – pi 54
- EE-1 : event name of Bt brinjal
- Pup-1 QTL improves phosphorous uptake in phosphorous deficient soils
- PSTOL 1 QTL: phosphorous starvation tolerance -1 enhances root system for P absorption.
- Pup-1 QTL , PSTOL 1 QTL, Sub-1 QTL's are isolated from kasalith rice.
- Pup-1 QTL absent in Nipponbare (japonica) reference genome of rice

- Naturally transgenic food crop – sweet potato – contains T-DNA of Agrobacterium which has evolutionary advantage because of expression of favourable genes in T-DNA.
- Sr -31 gene shows resistance to all races except ug99 race from Africa
- Sr -33 gene from Triticum monococcum is resistant to ug99 and some races but susceptible to few races.
- Sr -35 gene from Aegilops tauschii shows moderate resistance to all races including ug99 race.
- ROS scavenging enzymes : oxidases, catalases, super oxide dismutase
- Neofunctionalization – new gene from duplicated gene
- IRRI – 1960
- G. barbadense : 38mm fibre length, 32g/tex fibre strength
- G. hirsutum : 30mm fibre length, 30g/tex fibre strength
- SSR markers are polymorphic second generation markers
- World's best extra long staple variety – 'suvin' (G.barbadense)
- Fine and unique fibre quality - G. barbadense
- LRA 5166, Anjali, Suraj- CICR varieties
- MCU-5 : photoinsensitive mutant variety of G. hirsutum
- Intersubspecific sterility- indica x japonica crosses
- Chromosome 3B of wheat is double the size of rice genome ~400Mbp
- AAd gene is to select transformed bacterial cells from untransformed cells
- Improved hybrid bajra – HHB-67 (MAB for powdery mildew resistance)
- ICPH-8 : world's first pigeon pea hybrid
- ICPH 2671 (pushkal) : world's first CMS based pigeon pea hybrid
- Sha 1 gene responsible for shattering in rice
- Meg-1 : nourishing gene in maize from mother to offspring whch increase seed mass
- Flavr –savr : first GM in 1994
- International biodiversity year – 2010

- Indica – long grain types
- Japonica – short grain types
- First mustard hybrid – sankar sarson (NRC HB-506)
- FR13A (sub-1) transferred into swarna variety to produce 'swarna sub-1' by MAB
- Soyabean ferritin gene helps in high iron accumulation in endosperm than aleurone layer.
- Sd-1 gene was introgressed into indica types to produce dwarf types with retained grain quality.
- Tomato cv. Heinz1706 and its closest wild relative Solanum pimpinelli have been accomplished (genome sequenced) recently.
- HM-4 : baby corn hybrid
- HSC-1 : hybrid sweet corn
- First semidwarf indica variety in rice– TN-1
- Dee –geo-woo-gen (Taiwan) contains sd-1 gene
- Splikelet sterility – screening tool for heat tolerance in rice
- Samba mahsuri – GEB 24 x Mahsuri
- Swarna – vasistha x mahsuri
- Lignans in linseed act as anti oxidants
- IISR comp-1 : sugarbeet variety
- Enzyme Invertase plays major role in deterioration of sugarcane juice quality
- First double cross maize hybrid by shull – 1921
- True potato seed : quality seed with virus free
- Mega variety in wheat – PBW 343 with 1B/1R
- Deep rooting 1 (DRo 1) QTL control root growth angles in wheat
- First malt purpose variety in wheat – DWR UB-64.
- Oxalic acid and phytates – anti nutritional factors in sesame
- Ricin and agglutinin - anti nutritional factors in castor
- Castor – industrial use – eri silk rearing, ricinolic acid and ricin
- Sesame oil is highly stable because of γ tocopherol, sesaminol and sesmolin

- Linseed – high linolenic acid (70%)
- High oleic acid – sunflower and safflower
- Complementary epistasis – cross between two susceptible varieties appear to be resistant phenotype.
- Borlaug – outlined multilines
- Jenssen – proposed multiline in oat
- Gene deployement is a useful strategy to avoid boom burst cycle
- Parallel spindle fibre mutant- first meiotic division is followed by cytokinesis without undergoing second meiotic division results in dyad
- Registered seed in seed production chain involved in crops were low seed multiplication ratio.
- Chlorine : first commercial mutant of tobacco
- Minimum error degree of freedom in Augmented block design – 12.
- In latin square design – 4 sources of variation
- Type –I error : rejecting true null hypothesis
- Type –II error : accepting false null hypothesis
- Rejecting true null hypothesis : Type –I error (False positive) and Type –II error (False negative) in QTL mapping
- Sonora 64 was released for cultivation in the yellow rust free areas
- Lerma rojo 64 was released in yellow rust prone areas
- Genetic erosion : loss of wealth due to human intervention
- Genetic wipe out : loss of wealth due to natural disasters
- Base collection : 20yrs: 3 to 4% moisture level; -18 to -20°c
- Active collection : 10-15yrs: 8 % moisture level; 0°c
- Working collection : 3-5yrs: 8 to 10 % moisture level; 5 to 10°c
- Micro centre : narrow area in the centre of diversity
- Centre of origin : more of dominant alleles and contain wild spp
- Frequency of recessive alleles are high towards periphery of origin
- Crossing over between gene and centromere leads to reductional division in meiosis –II
- If there is no crossing over between gene and centromere then reductional division occurs at Meiosis –I itself.

- When the mutations are allelic or non allelic cis heterozygotes produce wild type
- Trans heterozygote produce wild type only when mutations are non allelic.
- $0 < r < 1$: partial dominance to complete dominance
- High heritability, high V_A – mass selection
- low heritability, high V_A – progeny selection
- $np_r = \frac{n!}{(n-r)!}$
- $nc_r = \frac{n!}{r!(n-r)!}$
- $nc_r.P^{n-r}.q^r$
- Incomplete penetrance and variable expressivity have a masking effect on the phenotype and therefore reduce the correspondence between genotype and phenotype
- Correlated response between two characters is due to linkage or pleitropy.
- Linkage relationship breaks during segregating generations but pleiotrophy remains.
- To study value of a gene, isogenic lines is promising approach even on different genetic background.
- Adaptive heterosis : true heterosis because it has role in fitness of individual
- Labile heterosis – not fixable
- Mean of equilibrium population = a(p-q) + 2pqd
- Inbreeding change = 2pqdF
- Ideotype – Bilological model which is expected to perform or behave in a predictable manner with in a defined environment.
- IS 11167 and IS 11758 – high lysine Ethiopian mutants in sorghum
- $\overline{Y_2} = \overline{Y_1} - \frac{(Y1-Yo)}{n}$ mean performance prediction of composites.
- Mass selection and bulk method are common in participatory plant breeding
- $F_n = (\frac{1}{2})^{n-1}H + M.P$: mean performance of any selfed generation.

- Residual heterosis in $HF_2 = 2(F_2\text{-M.P})$
- $H = 2\,\overline{BC_1} - \frac{1}{2}\,(3\overline{P_1}+\overline{P_2})$: Heterosis in B.C generation
- $H_{syn}\ \frac{n(\overline{Ysyn}-\overline{P})}{n-1}$:Heterosis in synthetics.
- $H_{com} = \frac{\overline{Ycom}-\overline{P}}{n-1}$ Heterosis in composites.
- Geographical indication – 1999 act.
- G.I tag – 10yrs
- Composite sample : true representative of seed lot
- PPV&FRA-2001, came into force in 2004 and came into effect from January 2005.
- Plant breeders right -1994
- Registration is not compulsory in G.I act
- Registration is compulsory in PPV&FRA
- Patent duration : 15-20yrs
- $V_E = \frac{VP1 + 2VF1 + VP2}{4}$
- Geometric mean = $\sqrt{(a \times b)}$
- S.D = $\sqrt{\text{Variance}}$
- S.E = $\frac{S.D}{\sqrt{N}}$
- Phenotypic expression of trait = $\mu + g + e + g \times e$
- F_2 genotypic ratio of 1 locus : 1:2:1
- F_2 genotypic ratio of 2 locus : 1:4:6:4:1
- F_2 genotypic ratio of 3 locus : 1:6:15:20:15:6:1
- Genotypic value $V_G = \frac{1}{2}\,V_A + \frac{1}{4}\,V_D$
- Breeding value = $\frac{1}{2}\,V_A$
- Dominance deviation = $\frac{1}{4}\,V_D$
- Co.V (H.S) is half of breeding value of parent i.e Co.V (H.S) = $\frac{1}{4}\,V_A$
- Co.V (F.S) = $\frac{1}{2}\,V_A + \frac{1}{4}\,V_D$
- Co.V (F.S) - 2Co.V (H.S) = $\frac{1}{4}\,V_D$
- $Co.V_{(op)} = \frac{1}{2}\,V_A$: open pollinated population
- $b_{(op)} = \frac{1}{2}\,\frac{VA}{VP}$: regression coefficient of open pollinated population

- $Co.V_{(op)} = \frac{1}{2} V_A$: mid parent progeny
- $b_{(op)} = \frac{VA}{VP}$: regression coefficient of mid parent progeny
- Genotypic advance = difference of mean genotypic values of selected lines and population.
- Line x Tester analysis : Vgca = Cov H.S, Vsca = Cov FS-2 Cov H.S; V_A = 4gca (F=0) , V_A = 2gca (F=1), V_D = 4sca (F=0), V_D = sca (F=1)
- In self pollinated crops if sca variance is 2 times to gca variance then only $V_A = V_D$
- $\frac{VA}{VD} > 1$: Additive gene effects are more
- $\frac{VA}{VD} >$ 1 : Dominance gene effects more
- NCD-I : $Vm = \frac{1}{4} V_D$, $Vf = \frac{1}{4} V_A + \frac{1}{4} V_D$: $V_A = 4Vm$, $V_D = 4Vf - 4Vm$
- NCD-II : $Vm = Vf = \frac{1}{4} V_A$, $Vmf = \frac{1}{4} V_D$
- NCD-III : $Vm = \frac{1}{4} V_A$, $Vmf = \frac{1}{2} V_D$
- Heat avoidance = dissipation of radiant energy, transpirational cooling, pubescence and glaucousness
- Freezing tolerance = membrane stability, osmotic adjustment, cold responsive proteins.
- In partial diallel no sca effects were estimated
- Number of sampled crosses in partial diallel = ns/2 ; s = n/2,
- Selection of sampled crosses in partial diallel : $K = \frac{n + 1 - s}{2}$
- Quadriallel analysis : 4 –line effects
- In genetic diversity analysis minimum D^2 value above which genotypes are selected in each cluster is $\frac{\varepsilon Di}{n(n-1)}$.
- Manifestation of phenotypic effects in the individuals which they contain dominant/recessive gene.
- Sexual isolation between male and female is isolation index
- Stabilizing selection favors additive gene action and occurs for the traits that do not affect fitness directly.
- Disruptive and directional selection favors dominanace and epistasis effects.

- Disruptive selection helps in establishment of integrated super genes.
- Linola in linseed contain 2 % linolenic acid.
- Threshold traits are basically multigenic but expression is like discontinuous
- Heterozygous balance : hides unfavourable genes to avoid detrimental effects.
- TGMS in rice : < 23.3°c – male fertile; < 23.3°c – male sterile.
- Soyabean oil contains high linoleic fatty acid.
- Golden rice –I : 8g/g of provitamin A levels - 2001
- Golden rice –II : 36.7g/g of provitamin A levels - 2005
- 'Bar' gene shows resistance to herbicide Phosphinothricin
- 'Hpt' gene shows resistance to hygromycin antibiotic.
- Recessive mutants of Fad 2 gene in soyabean helps in increase accumulation of oleic acid
- Soyabean ferritin gene, metallothionein and Aspergillus Phytase gene increases iron availability and absorption.
- Good substitute for sesame in oilseed crops – Niger because of high linoleic acid
- Undesirable components in sunflower oil are chlorogenic acid and caffeic acid
- Undesirable components in rapeseed oil are erucic acid (22:1) and eicosenoic acid (20:1)
- Balanced oil quality with high oleic and linoleic ratio – Virginia runners in groundnut
- High oil content – Spanish bunch types in groundnut
- Protein component in groundnut is Arachin
- Sunflower oil contains high oleic acid and high vitamin A content which makes more stable.
- Oldest oilseed crop is sesame
- Traitor gene technology contains inducible promoter – to have expression of particular gene of interest inducer is essential to silence repressor molecule.

- Terminator gene technology – contains 'cre-lox' mechanism which makes seeds to germinate for one generation only.
- Polyunsaturated fatty acids contain more than one c=c bonds
- Free fatty acid < 2 % : oil quality will be poor
- Oleic group : omega -9 fatty acid
- Linoleic group : omega -6 fatty acid – Gamma linoleic acid, arachidonic acid and eicosadienoic acid (20:2)
- Linolenic group : omega -3 fatty acid – alpha linolenic acid (18:3), Dicosahexaconidic acid (22:6)
- Elongase enzyme increases conjugated fatty acids
- Lipoxygenase removes c=c bonds.
- High density lipids – PUFA – good for health but high proportion deteriorates oil quality.
- Parent offspring regression : heritability (H)=2b
- ABA inhibits mRNA and protein synthesis during water stress condition
- H.I in cereals : 0.4-0.6
- H.I in pulses : 0.3-0.4
- H.I in oilseeds : 0.3-0.4
- Nitrogen H.I: ratio of nitrogen in grain to the total nitrogen content of plant biomass.
- To alter gene frequency in desirable direction cyclical/recurrent selection are useful.
- Harvest plus programme was started by Howarth bouis – 2003.
- Food prize for Harvest plus programme - 2016
- CEL I endonuclease enzyme cleaves mismatches in heteroduplex analysis of TILLING.
- TILLING: SNP detection, Marker development and gene function.
- ECOTILLING : study of existing natural variation across germplasm lines in a particular locus.
- RNA I : Fire and Mello
- Raffinose – Antinutritional factor that causes flatulence in chick pea
- PNLOX 13S gene blocks synthesis of aflatoxin in groundnut.

- Dhansakthi (ICMH-1201) variety in bajra contain high Fe and Zn content.
- GM mustard – DMH-11
- ICCV 2 – world's shortest duration chick pea variety
- Designing superior genotypes insilico is ' breeding by design'
- Yield stress index = $\frac{\text{Yield under stress condition}}{\text{Yield under normal condition}}$
- TASSEL – Trait analysis by association, evolution and linkage
- Mutation, recombination and gene conversion reduces linkage disequilibrium
- Lysimeter – evapotranspiration
- Neutral markers – differences in DNA sequences that do not confer a selective advantage/disadvantage during QTL mapping.
- Ma^R : photoperiod insensitive maturity allele in sorghum
- 'Stock 6' – founder for haploid inducer gene stock in maize.
- Anthocyanin color marker – R_1 –Navajo (nj) – a dominant marker which express color in scutellum of embryo and aleurone layer of endosperm helps in detection of haploids in maize.
- Morel and martin – meristem culture.
- Primer 3: data base to find oligonucleotide primers for PCR amplification of DNA sequence.
- Primer blast: finding primers specific to PCR template.
- Primer X : automated design of mutagenic primers for site directed mutagenesis.
- RNAi design : duplexed RNA oligo's for RNA interference.
- Annealing temperature $<$ melting temperature – primer will not anneal
- Annealing temperature < melting temperature – mismatch annealing
- Melting temperature of primers should be around 55 to 72°c.
- Expressed sequence tags – 200 to 300bp long cDNA sequences
- TBLAST N – translate DNA databases
- BLAST X – translate input sequences

- TBLAST X – translate DNA databases
- Vectors like BAC, YAC and PAC are used to construct gene libraries because they can take even 1 Mbp DNA fragment.
- Allele mining – dissect naturally occurring genetic variation of useful agronomic traits
- Poly cistronic transgene : single transcription unit with regulatory sequences is to express several effect genes from a single promoter
- Boll gard –II : cry 1 Ac + cry 1Ab : resistance to helicoverpa and spodoptera
- Complete set of chromosomal and extra chromosomal genes present in and organism.
- Golden rice –I : contain 3 genes (2 from daffodil and 1 from bacteria)
- Golden rice –II contains genes from maize and bacteria
- Allele mining is based on candidate gen approach or sequence based
- Hordeum Vulgare sub spp spontaneum – wild ancestor of barley
- Non-shattering domesticated type developed by a mutation in one of two tightly linked genes Bt_1 and Bt_2 in barley.
- Domesticated barley types are 6 rowed types unlike wild 2 rowed types
- Genome sequenced in barley – 2012
- Malting barley contain low protein content and thus more fermentable sugar content.
- Rice monophyletic origin from oryza perennis
- Saline water tolerant variety in rice– Lunishri
- GEB 24 and basmati 370 are pureline varieties of rice
- IR-8 : miracle rice
- Chinsura Boro-II : CMS source of indica background in rice
- WA cytoplasm : CMS source of japonica background in rice
- Salt tolerance : exclusion of salts, osmotic adjustment, accumulation of salts in roots itself, selective accumulation of 'K'.
- Xa_{21} and Xa_4 :these genes shows moderate resistance to blight in rice in singly also.
- Xa_5 and Xa_{13} – ineffective singly, so combination is used.

- Bph-1 :first identified gene in rice
- World's first dwarf basmati variety – pusa basmati -1 (TN-1 x basmati 370)
- Late embryogenesis abundant protein – salt and drought tolerance in most of crops
- Proline synthesis by enzyme Pyrroline -5-carboxylase synthetase
- Population improvement: broadens genetic base of a population, increase of frequency of desirable alleles, breaks linkages and increases buffering capacity.
- Basmati features : extra kernel elongation and fragrance and least swelling
- TGMS : "Annong-IS", PGMS- japonica -"Nong-Ken 58" in rice
- Shuttle breeding : generation and selection of population at two different locations : increases stability.
- Jalapriya, jitendra, nalini : deep and semi deep water varieties in rice
- GMS in sunflower was first identified by "Kuptsov"
- Sporophytic self incompatibility – main mechanism in operating cross pollination in sunflower.
- Eric D.Putt- exploited the heterosis in sunflower.
- CMS in sunflower was identified by "Leclercq"
- Pustovit's method of reserves is a method of recurrent selection in sunflower.
- Sugarcane is a complex polyploid with basic chromosome no -8
- Saccharum spontaneum – wild ancestor of sugarcane
- Co 205 (3n)– S. officinarum (2n) x S. spontaneum (n)
- Nobilization of cane : introgression of genes from wild spp in to cultivated spp (S. officinarum)
- Tifton 23A cytoplasm of bajra is susceptible to downy mildew
- Athwal – bajra breeder
- Bajra used as forage crop because of low HCN content.
- Intergeneric hybridization with Erianthus, miscanthus helps to improve juice quality.

- Sugarcane is not ideal to study genetic studies because of multivalent formation, complex polyploid, aneuploid gamete formation.
- Superior juice quality is in S. officinarum
- True seeds of sugarcane : fuzz
- Proven parent in sugarcane: high proportion of desirable individuals in progeny
- Proven cross in sugarcane : progeny consists of high frequency of desired recombinants or clones of good commercial merit.
- Proven parent need not be proven cross
- Poly cross / melting pot method in sugarcane: random mating between clones.
- Indian canes : Saccharum barberi – tolerant to low temperatures
- Saccharum spontaneum – tolerant to biotic and abiotic stresses in sugarcane.
- Large number of seeds has to be conserved in a crops like sugarcane because of highly heterozygous nature.
- Commercial wonder variety in sugarcane – CoJ-64
- Infra red spectroscopy – sugar quality
- Triticum timopheevi – 4x –AAGG
- Necrotic genes in wheat which necrosis during combinations – Ne-1, Ne-2
- CMS source in wheat - Triticum timopheevi
- DDK 1001 and DDK 1009 were first semidwarf dicoccum wheat in world developed in India.
- In India monosomic series in wheat were developed in variety kalyansona.
- Ph^1 gene in long arm of 5B chromosome prevents homeologous pairing.
- Kharchia local shows salinity tolerance in wheat.
- The relationship between protein content and lysine content in cereals is negative.
- Adult plant resistance : horizontal resistance which keeps infection levels low.

- Seedling resistance : vertical resistance (r = 1)
- Non – enzymatic storage protein in wheat – gluten
- Gene cycling ensures temporal delay in the use of a resistance gene. This discontinuity minimizes the evolution of complex virulence and increases the variety of vertical resistance genes.
- A minor component of endosperm rich in lysine is "Triticin".
- Secondary gene pool of wheat contain spp of Triticum and Aegilops which contain one genome in common.
- Wild spp which contain apomictic genes in pearl millet is Pennisetum purpureum
- Lr 10 gene shows resistance to race 77.
- Teosinte – ancestor of present day maize.
- Masculinizing genes : tassel seed 2(ts 2) required for gynoecial abortion in maize and feminizing genes : anther ear 1 (An 1) and silkless (sk 1) arrest stamen development.
- Composites have better advantage to survive under less productive soil and drought conditions rather than the high input required hybrids.
- Hybrids between maize and tripsacum are partially female fertile and complete male sterile.
- CMS (C) : charrua cytoplasm, CMS (S) : USDA cytoplasm and CMS (T) : Texas cytoplasm are different CMS sources in maize
- CMS (T) : Texas cytoplasm is commonly used because of full sterility and easy restoration but susceptible leaf blight.
- Multiplasmic line : combination of different kinds of sterile cytoplasm
- Jenkins : prediction of double cross
- Heterotic pool : broad based, genetically diverse, highly heterotic, inbreeding tolerance and acceptable per se performance.
- $\frac{n(n-1)(n-2)}{2}$: three way cross
- $\frac{n(n-1)(n-2)(n-3)}{8}$: double cross

- Apomictic genes are transferred into maize from Tripsacum dactyloides
- 1 generation of selfing = 3 generations of Full sibmating = 6 generations of Half sib mating = 1 genration of back crossing with a homozygous parent : rate of inbreeding
- Composites in maize : vijay, sona, amber, kisan, jawahar, vikram, ageti-76, parbhat
- First single cross maize hybrid in India : 1957
- Histarch, Trishulatha, Ganga 4, Ganga 5, Ganga 11, Deccan 105 : Three way cross hybrids.
- Ganga 1, Deccan, Ganga 3, Sartaj : double cross hybrids.
- Gossypium herbaceum varieties : jayadhar, digvijay
- Gossypium barbadense : sea island/pima cotton/Egyptian cotton
- Secondary gene pool of cotton contains A, B, D and F genomes of diploid spp
- Gossypol : Antibiosis towards pest tolearance in cotton.
- Asiatic diploid spp used in coarse fabric manufactures
- Selection of two extreme types for a character in segregating generation and allow intermating between them – Disruptive selection.
- Lakshmi x SB 289E – vara lakshmi (4x - inter specific hybrid) had high fibre length in cotton
- DCH-32 (4x- inter specific hybrid) in cotton had high fibre fineness and strength
- Intra hirsutum hybrids in cotton : PHV Hy 2, H-4 and savitha
- Biparental mating in segregating generations is used to break undesirable linkages and to release variability.
- Progeny test : discriminates homozygous/heterozygous mother plants and to know superiority of plant is due to genotype or environment
- For hybrid development male parent is used as facultative apomictic to transfer genes for apomixes into progeny. Selection of stable obligate apomictic plant in F_2 generation has to be carried out to be released as vybrid.

- Linked gene hypothesis in dominance hypothesis of heterosis is by jones
- Explanation for continuous variation for few traits in dominance hypothesis of heterosis is given by Collins.
- Convergent improvement : simultaneous improvement of both lines in back cross breeding.
- Double cross hybrids have high genetic base than single cross hybrids because of high heterogeneity.
- Primary gene pool is useful for transfer of desirable alleles and broadening of genetic base of breeding population.
- Secondary and tertiary gene pools are used for creating vast genetic variability for various oligogenic and polygenic traits.
- Powerful tool for comparative and synteny mapping : RFLP
- SSR's – multiallelic, SNP – Biallelic
- Soyabean cyst nematode resistance gene : rhg 1
- Williams et al – RAPD- dominant marker
- Vos et al – AFLP – dominant marker
- Co-dominant marker –can distinguish homozygotes and heterozygotes
- Dominant marker – can't distinguish homozygotes and heterozygotes
- Locus specific marker : RFLP, RAPD, SSR, SNP
- Locus non specific marker: AFLP
- Marker assisted selection is highly useful for traits with low heritability
- The term isogenic coined by johannsen
- Composite interval mapping – estimates the effects of the QTL by removing the effects of other QTL's linked to the locus.
- Multiple interval mapping – simultaneously to fit multiple QTL directly in the model for mapping QTL and also determines interaction of significant QTL's and their contribution to the genetic variance.
- 6-10 NIL's are mixed together to constitute a multiline cultivar.
- 5-8 best general combining inbreds are used to constitute synthetic variety.

- Progeny produced through continuous selfing in cross pollinated crops is called inbred line.
- Progeny of single self fertilized homozygous plant is called is pureline.
- Single locus fingerprinting : polymorphism at a single locus is analysed through DNA probe/specific PCR primers.
- Multi locus fingerprinting : single probe that identifies multiple similar sequences.
- The maximum number of transgenic cultivars have been released in rapeseed followed by potato in world.
- Protein quality in alfalfa : transgene ovalbumin gene of chicken
- Orange flower color in petunia: transgene from corn.
- Anti sense RNA technology : 'Flavr savr' in tomato for poly galactouronase gene
- Anti sense RNA technology : 'Endless summer' slow ripening tomato by inhibition of ACC synthase.
- Particle bombardment method : no control over copy number and site of integration of gene.
- First genome sequenced organism – Haemophilus influenza
- Varieties which are selected and cultivated by farmers for many generations : primitive cultivars
- Improved variety of past : obsolete cultivar
- A variety which is used for comparing a particular trait in DUS testing : Example variety
- Smallest genome size : Mycoplasma genitalium (0.58Mbp)
- Study of all metabolic pathways in a living organism : metabolomics
- The seed that has been multiplied under organic conditions atleast for one generation is referred to as organic seed
- The cultivars that originate from organic plant breeding techniques are known as organic varieties.
- Participatory variety selection: selection within stable (homozygous) lines or populations involving farmers.
- Pariticipatory plant breeding deals with segregating material i.e F_3 onwards of a cross.

- In traditional breeding first the variety is released and then adopted by farmers.
- In participatory plant breeding, the new varieties are first adopted by the farmers and then released.
- Usually mass selection and bulk breeding methods are followed in participatory plant breeding.
- High parent progeny regression and low mean deviation of offspring from the population mean gives high breeding value.
- Adaptation : process of adjustment of living organisms to the changes occurring in the environment.
- Suitability of a variety for general cultivation over wide range of environments is stability.
- Specific genotypic adaptation Eg: high internode length in deepwater area of rice
- Specific genotypic adaptation Eg: semi dwarf varieties of rice and wheat
- General population adaptation Eg: composites
- Specific population adaptation Eg: synthetics
- sucrose (su), shrunken (sh), amylose extender (ae) – sweet corn mutants for different loci.
- The less Anthesis – silking interval the more drought tolerance in maize under stress conditions.
- Actual green revolution - 1970
- Genetic erosion: gradual reduction in genetic diversity in the population of a species due to elimination of various genotypes.
- Mass selected variety : mixture of purelines in self pollinated crops : homozygous heterogenous population
- Synthetics and composites : mixture of homozygotes and heterozygotes
- The 5B system in wheat was discovered by Riley in 1954.
- The commercial potato is an autotetraploid though interspecific hybridization also involved.
- Genetic causes for deterioration of variety – adaptive changes, residual variations, mutations, cross pollination

- Non genetic causes for deterioration of variety – mechanical mixtures, disease infestation, adverse agroclimatic conditions
- Diploid lines which are obtained by doubling the chromosome number of a haploid line – doubled haploid
- A haploid individual which is produced from a tetraploid spp is referred as dihaploid.
- A haploid individual with haploid chromosome sets of two diploid spp is referred as double haploid
- A variety that has to be registered under PVP act – candidate variety
- All released and notified extant varieties that are in seed production chain – Reference variety
- A variety that has been developed by a farmer and used for commercial cultivation for several years.- farmers variety
- All released and notified varieties which have not been protected – Extant variety
- The genetic polymorphism which is maintained due to superiority of heterozygote over the both homozygotes is referred as balanced polymorphism.
- Variety should not be in public domain for more than 1 year is provided plant variety protection.
- Broad genetic base is nothing but mixture of homozygotes and homozygotes in population which has high buffering capacity towards variable environments.
- Doubled haploid lines often appear inferior to inbred lines obtained by conventional method because of very high level of homozygosity.
- Brassica juncea – AABB
- Breakage fusion bridge cycle initiated by telomeric dysfunction giving rise to unstable dicentric or ring chromosome.
- Interstitial deletion gives acentric fragments which eventually lost from a cell population.
- Complex translocation – more than 2 breaks and more than 2 chromosomes involved in exchange.
- Pseudo dicentric : one centromere is inactivated in dicentric and behaves as monocentric.

- There are no consequences if crossing over occurs outside the inversion loop in paracentric inversions.
- When there is crossing over within the inversion loop 50 % of gametes are sterile because of deletions and duplications with acentric and dicentric chromosomes.
- Transverse division of centromere in chromosome during mitosis gives isochromosome formation.
- Concept of heterotic pool was developed in maize.
- Lr 2, Sr 9 provide durable resistancee in wheat.
- Vertifolia effect in potato : erosion of horizontal resistance in the process of continued breeding for vertical resistance.
- Lr 19 and Lr 28 : Aegilops elongatum
- Lr 9 and Lr 24 : Aegilops umbellulata
- Neomorphic mutation : gain of function
- Amorphic function : loss of function
- Neutral mutations : these mutations do not have effect on fitness of spp.
- INDELS are irreversible mutations
- Mutation rate is high in RNA than DNA because RNA replicates do not have proof reading activity.
- Base analogs cause mutations in replicating DNA only
- Alkylating agents cause mutations in replicating and non replicating DNA also.
- Uv rays results in formation of photodimers on the same strand
- Photolyase breaks dimers into monomers
- INDELS occur at repetitive sequences of DNA
- Apurination and deamination are spontaneous mutations.
- Uracil DNA glycosylases carry out DNA excision repair system
- Sister chromatids replicates continuously without cell division so that numerous strands were formed and joined at the centromere – Endoduplication
- If one such strand separates, it acts as individual chromosome which leads to condition – Endopolyploidy

- Number of giant chromosomes in salivary glands of drosophila = n
- The form, size and number of chromosomes in the basic set of an individual constitutes its karyotype
- Arrangement of chromosomes in order of decreasing mean length from left to right placing the long arm downwards - Idiogram
- Chiasmata is visible during diplotene stage when homologous chromosomes move away from each other.
- Classical theory – chiasmata is cause of crossing over and each chiasmata does not lead to crossing over.
- Chiasmata theory – chiasmata is consequence of crossing over. Chiasmata appears at the point where crossing over has taken place.
- Chromosome banding – Area/region/sections of chromosomes which stains more deeply than the neighbouring parts as a result of chromosome treatment with dyes.
- C – banding – Giemsa stain – constitutive heterochromatin
- Q – banding – Quinacrine stain –
- G – banding – Giemsa stain –
- N– banding – Giemsa stain –
- Single DNA double helix is complexed with histones to form a strand/ chromonema commonly known as – Chromatin fibre
- Knob like structure appears near to secondary constriction - Satellite
- Chromosome painting – Insitu hybridization with probes labelled with specific fluorescent labels for each chromosome.
- Genome blocking – Detection of alien chromosome or chromosome segments on wheat background by adding alien chromosome along with wheat genome, so that alien chromosome segment pairs with alien chromosome.
- Folded fibre model by dupraw explains the chromosome structure.
- Nucleosome solenoid fibre model – Kornberg and Thomas
- Sister chromatids twisted round each other – Plectonemic coiling in Mitosis
- Sister chromatids slipped over each other – Paranemic coiling in Meiosis I

- There is no relational coiling in Meiosis –II
- Formation of multipolar spindle formations leads reduction in chromosome number in a cell during mitosis – Somatic reduction
- Diakinesis – end of chiasmata terminalisation, spindle fibre organization and nucleolus disappears.
- DNA synthesis took place and mitosis was initiated but aborted before anaphase due to failure of spindle fibre formation which eventually lead to return to interphase state and results in formation of restitution nuclei with double chromosome number – Endomitosis
- Somatic crossing over in drosophila – Stern
- Chromosomal fibres – spindle fibre origin at one pole and move to centromere during cell division.
- Two Acrocentric chromosomes at their short arms fuse to form microchromosome and metacentric chromosome – Roberstonian translocation.
- Change in position of centromere because of pericentric inversions when the two breaks are not equidistant to the centromere.
- Cross over products in paracentric inversion contain acentric and dicentric chromosomes having deletions and duplications which leads to pollen sterility.
- Cross over products in pericentric inversion contain monocentric chromosomes having deletions and duplications which leads to pollen sterility.
- Translocation heterozygotes gives 50 to 60% pollen sterility because functional gametes were produced only by alternate segregation rather than Adjacent –I and Adjacent –II which contain deletions and duplications.
- Inversions – Sturtevant
- Deletions, Duplications and Translocations – Bridges
- Balanced lethal system in drosophila – Muller
- Permanent hybridity at gametic and zygotic level is maintained in translocation heterozygote – Balanced lethal system.
- Balanced lethal system is because of repulsion phase with recessive lethal condition where homozygous lethal do not survive which leads to retaining of heterozygosity only.

- Autosyndetic pairing – pairing between chromosomes derived from the same parent.
- 'hap' gene induces haploidy in barley.
- 'ig' gene induces androgenetic and gynogenetic haploids in barley.
- Conditions in Chromatid segregation – Multivalent formation and homologues carrying cross over product move to same pole.
- Triallel and quadriallel analysis provides information on epistatic variance unlike diallel analysis.
- Line x Tester analysis – 50 lines
- Diallel analysis – 10-12 lines
- Partial diallel - 20 lines
- Linear combination of traits associated with yield – Selection index.
- Microsatellite - 2 to 7 bp short tandem repeats
- Minisatellite - 15-60 bp short tandem repeats
- Physical map – Distance between two genes interms of base pairs separating them
- Ganga – First hybrid maize
- Father of cytogenetics – Muller
- Bateson – coined terms Allelomorphs, Homozygote, heterozygote
- Crops introduced by portugese into India – Maize, groundnut and potato
- Crops introduced by East India company into India – Tea, Litchi and Cabbage
- Earlier name of NBPGR – plant introduction and exploration organisation
- Hybrid necrosis in wheat is avoided when grown at temperature ↑
- Mating between individuals of less closely related by ancestry – Genetic disassortative mating.
- Progeny obtained from self fertilization were weaker than those derived from outcrossing – 'Darwin'
- Alfalfa – High inbreeding depression
- Maize, bajra and sorghum – Moderate inbreeding depression

- Germplasm conversion – Modification of germplasm into utilizable form otherwise in unutilizable form.
- Pedigree - 13-14 years
- Mass selection – 5 to 7 years
- Pureline variety – 9 to 10 years
- Spontaneous mutation frequency – 10^{-4} to 10^{-6}
- Convergent cross – More than two parents are crossed to produce the hybrid
- Presently the centres of crop origin are – 12
- Substations of NBPGR – 5
- Low temperature treatment of 0-6 ^{0}c for 15-24 hrs leads to pollen sterility in rice
- First patent was awarded for the gene 'aro A' isolated from mutant bacterial strain and intended into plants to confer glyphosate resistance.
- Indian patent act – 1970
- International plant Genetic Resource Institute – Biodiversity international
- Plant breeder rights - 15 to 20 years
- Indian Patent office – kolkatta
- Plant variety protection – 1970
- National biodiversity authority – Chennai - 2002
- IR8 : miracle rice
- IR8 : susceptible to bacterial blight
- Cartagenea protocol : biodiversity
- Super cooling is a mechanism of freezing avoidance
- Winter hardiness : rye↑ wheat↑ barley↑oats
- Antifreeze gene chemically synthesized – 'ala3'
- Meiosis in anthers is the most susceptible to chilling
- High proline accumulation – chilling tolerance
- Drought avoidance – water savers and water spenders
- Water savers: leaf rolling, senescence, waxiness

- Water spenders : high water uptake : deep root system, root length density
- ABA accumulation increase chill tolerance but not freeze tolerance
- Sub-1 gene transferred into MTU-7029 (swarna) for submergence tolerance
- Heat avoidance : Transpirational cooling
- Convention on biological diversity : 1992
- Agra local : susceptible to all types of rust
- Vertical resistance : for soil borne pathogens
- Rate of disease reaction for avirulent strain in vertical resistance is zero
- Rate of disease reaction for virulent strain in vertical resistance is one
- Vertifolia effect given by Vanderplank
- Salicylic acid accumulation gives systemic acquired resistance
- Horizontal resistance : for air borne pathogens
- Hypersensitive reaction is for obligate parasites
- Tift 23A male sterile cytoplasm is susceptible to downy mildew
- Canola - ↓ 2% erucic acid and low glucosinolates
- Rapeseed and mustard – single zero varieties : low erucic acid
- Double zero varieties : low erucic acid and low glucosinolate
- Triple zero varieties : low erucic acid, low glucosinolate and low fibre
- Morphological or anatomical resistance : durable resistance
- Non preference/Non-acceptance/Antixenosis
- Cabbage aphid – low sinigrin content
- Spotted cucumber beetle – low cucurbitacins
- Pea aphid – low amino acid and high sugar contents
- Spotted alfalfa aphid – high saponin content
- Rice stem borer – high silica content, lignified stem, sclerenchymatous tissue
- Brown plant hopper in rice – low asparagine content

- European corn borer – DIMBOA
- Maize stem borer – High asparatic acid
- Corn leaf aphid – low hydroxamic acid
- Cotton jassids – hairiness of leaves – Non preference
- Wheat green bugs – high content of benzyl alcohol
- Mung bean : anti nutritional factors : raffinose and stachyose
- Grain elongation on cooking : superior trait of basmati rice
- Polshenke value : <150 min – bread making
- Polshenke value : 80-150 min – chapatti
- Polshenke value : <80 min – Biscuits and confectionary
- Cotton halo length – length of fibre with attached seed
- Fibre fineness – micronaire value
- Indica rice varieties have high amylose content
- Fibre strength measured by stelometer
- Major part of genetic variation for fibre quality in tetraploid cotton is due to their 'D' genome contributed by a diploid progenitor
- Perfect aminoacid balance : pulses and cereals of 1:3 ratio for diet
- Essential aminoacids – 8
- High lysine mutants in barley – Hiproly, Notch-1 & 2, Riso 1508
- High lysine mutants in sorghum – P721, IS 11167, IS 11758
- Pusa red plum – Lycopersicum esculentum x Lycopersicum pimpinellifolium
- Albumins – water soluble
- Globulins – dilute saline soluble
- Prolamins – alcohol soluble
- Glutelins – Dilute alkali/acid soluble
- Legumes – deficient in sulphur containing Amino acid
- Cereals – deficient in lysine and tryptophan
- Anti nutritional factor : Haemaglutinin – broad bean
- Anti nutritional factor : phenolic glycoside – groundnut skin
- Anti nutritional factor : cyanogenic glycoside – forage legumes

- Anti nutritional factor : phytic acid – pearl millet
- Anti nutritional factor : Trypsin inhibitor – soyabean
- Major protein fraction in maize – Zein
- Wheat high protein genotypes – Atlas and Naphal
- Major protein fraction in pea – Vicilin
- Free variability : phenotypic differences between homozygotes with extreme phenotypes
- Genetic drift leads to loss or fixation of gene i.e 0 or 1
- Transgressive segregants appear in F_2 population because of polygenes
- Multiple correlation is non-negative i.e from 0 to 1
- The sum of square of deviations of all the observations of a samples from its mean is standard deviation
- The value of regression and correlation is the same when there is perfect correlation between two variables.
- Correlation is independent of measurement

Important Points on Seed Science & Technology

- Tag colour of Breeder seed is golden yellow
- Isolation distance for foundation seed production of sunflower hybrids is 600
- The chemical used for enclosing the synthetic seed is Sodium alginate
- Acid recommended for breaking physical dormancy is Sulphuric acid
- Convenient method of seed collection in teak is ground collection
- Example for Monocarpic plant is Bamboo
- The species which required after ripening period for dormancy breaking is Ginkgo
- April-May is the best suited season of planting for TPS
- The fruits which passed through the animal intestine and aids in seed dispersal is called as Endozoochory.
- Isolation distance for foundation seed production of onion is1000m
- In wheat, grain dormancy is influenced by environment
- Optimum season of planting onion bulbs for seed production is Nov-Dec
- Colored pigment present in immature cotton seed is Gossypol
- Progeny of nucleus seed is breeder seed
- Seed Village concept is adopted in Andhra Pradesh and Karnataka
- The genetic tool utilized for hybrid seed production in Brinjal is Manual modification
- MCU-5 cotton is a essentially derived variety (EDV)
- The headquarters of Central Institute for Cotton Research (CICR),is located at Nagpur
- The breeding tool utilized for hybrid seed production in redgram is GMS
- A pseudocereal crop is amaranthus

- The environmental factor which influences the seed dormancy in wheat is temperature
- In potato, sprouting is arrested by spraying MH
- The inflorescence of onion is umbel
- Presence of Embryo less seed is common feature in carrot
- The orchid seeds are germinated in nutrient agar medium
- The examples for the seeds without nutritive tissue (endosperm) is orchid seed
- Seed is a Matured ovule, Embryo enclosed in a seed coat and any propagative material
- The chemical responsible for dormancy in seed is Absissic acid
- The plant pigment responsible for seed dormancy and seed germination is Phytochrome
- The dormancy present in seeds of crops belonging to Fabaceae family is Physical dormancy
- The characteristic feature of morphological dormancy is the presence of Under developed embryo
- The eminent feature that differentiate Angiosperm and Gymnosperm seed is Seed coat
- The synthetic seed is Artificial seed
- The chemical used for enclosing the synthetic seed is Sodium alginate
- The modern techniques for varietal identification is Electrophoresis
- The enzyme responsible for prediction of viability through Tetrazolium test is Peroxidase
- The book on Seed Viability was written by E. H. Roberts
- Weed seed, other crop seed and inert matter are the fractions of purity analysis
- In seed quality control system golden yellow tag is given to indicate Breeder seed
- The color of seed tag for Foundation seeds is White color
- The color of seed tag for certified seeds is Blue color
- As per seed certification procedures the tag should be accompanied with producer label

- In India, the total number of State Farm Corporations operated are Two
- The seed control order emphasizing the need for License for sale of seed
- The New Seed Policy was introduced in seed industry during 1988
- In India Seed certification is Voluntary
- Indian seed certification rules are based on the rules proposed by OECD
- Certified seed is the progeny of Foundation seed
- Azar blue color tag is enclosed with the seeds bag of Certified seed
- The designated disease of wheat is Loose smut
- Dodder is the objectionable weed plant in Lucerne
- The media used for bitter gourd seed germination test is Sand
- Days for first and final count for paddy seed germination test is 5 and 14 respectively
- Certification could be done only for Notified varieties
- Seed Act was formulated and enacted during 1966 and 1968 respectively
- Seeds less than half size, during purity analysis is considered as inert matter
- Diaphono scope is used in purity analysis
- The working sample size for groundnut is 1000gm
- Phenol color test is used for identification of Varietal identification
- Grow out test is practiced to ensure genetic purity
- ISTA was established during 1924
- Fusarium wilt in Cicer orietanum is spread through Seed
- KOH bleach test in sorghum is useful to assess Varietal purity
- Sodium hydroxide test for wheat is useful in assessing Varietal purity
- Peroxidase test for soybean is useful to assess Varietal purity
- Phenol test for wheat is useful to assess Varietal purity
- Nobbe trier is useful in Sampling

- In cotton number of plants/count in field inspection is 250
- The submitted sample size for paddy is 400 g
- The RH required in a germination room is 90 %
- Electrophoresis is a test useful for Varietal identification
- Minimum weight of submitted sample for Moisture estimation in paddy is 50 g
- Boerner divider is otherwise called as conical divider
- Number of primary sample to be taken from the paddy seeds stored in 35 containers is 10
- Number of primary sample to be taken from 4500 kg of sorghum seeds stored in bulk is 7
- The working sample size for tomato is 7g
- Working sample size for maize is 1000 gm
- Working sample size for groundnut is1000 gm
- In Coniferae, seed without seed coat is considered as inert matter
- Genetic monogerm cultivars of beet are considered as Mericarp
- Number of seeds to be tested for assessing germination percentage as per ISTA is 400
- In estimation of moisture content the container used for seed packing should be moisture vapour proof
- Brick gravel test is a Vigor test
- Brick gravel test is used to measure physiological stamina
- Speed of germination is physiological vigor test
- Tetrazolium test is also a Biochemical vigor test
- Number of counts required for a maize field of 5 ha is 9
- Central Seed Testing Laboratory is located at IARI, New Delhi
- Central Seed Testing Laboratory was established during 1960
- Validity period of fresh seeds as per certification standards is 9 months
- International Seed Network Initiative was created by FS
- International Seed Network Initiative was started during June 1999
- Section 5 of Seeds Act deals with Notification

- The size of the certification tag is 15 X 7.5 cm
- Important seed borne disease accounted in seed certification of wheat is loose smut
- Grow out test is must for hybrids of cotton
- Seeds (control) order is proposed during 1983
- Designated disease of Knol khol is Blackleg
- Minimum germination required for cauliflower seed as per Minimum Seed Certification Standard is 65 %
- A test to determine field emergence potential is Brick gravel test
- Foundation seed class is certified by the Seed Certification Agency
- Minimum germination percentage recommended by Minimum Seed Certification Standards for tomato is 70 %
- Seed Act was implemented in India from 29th Dec 1966
- Seed rule was implemented in India from 2nd October 1969
- Plant quarantine act was passed during the year 1976
- Phytosanitary certificate assures that the seed is free from pathogens
- Plant Quarantine regulations was passed during the year 1981
- Minimum germination standard for paddy foundation seed 80 %
- The pure seed fraction recommended for groundnut is 99 %
- In Seed Testing Laboratory dividers are used to obtain working sample
- The isolation distance recommended in meters to avoid contamination with Johnson grass in hybrid sorghum is 400
- Pearson square law is adopted for Blending
- Objectionable weed plant of cucumber is Cucumis hardwickii
- Indian Seed Certification Rules are based on the rules proposed by OECD
- Without Phytosanitary certificate for smut disease the onion seeds cannot be exported to Pakistan
- Isolation distance for foundation seed production of onion is 1000 m
- In certification, the number of counts to be taken from 6 to 10 acres is 6

- In seed testing, permitted deviation from the standard is called Tolerance
- The designated disease of sorghum is Grain smut
- Pod verification is done in Groundnut
- The sample taken at one point in a seed lot is primary sample
- Cob sorting is done to identify off type cob
- Validation period for the information in certification tag is 9 months
- The seed samples sent for analysis by the seed inspector is official sample
- As per ISTA the germination test has to be conducted with pure seed fraction
- Notification is done under seed act to bring the variety to bring under purview of seed certification
- Seed sampling in cotton is done using hand sampling
- The submitted sample size for amaranthus is 1g
- Seed testing is the science of evaluating seed for its planting value
- Referee testing of seeds is being conducted at central seed testing lab
- In rice KOH test is useful for varietal identification
- As per ISTA working sample should contain a minimum of 2500 seeds
- In Tetrazolium test the viability of the seed is assessed by the development of pink color on all viable portion
- The maximum area could be registered for seed certification is 10 ha
- Image analysis of seed for varietal identification is based on morphological features
- Validation for a seed lot can be obtained for Any number of times
- To provide alternate temperature in a germination test, the high temperature is to be followed for 8 hr
- The light intensity required for germination of non-dormant seed is lux 250
- Seed testing is to be done under controlled condition
- Germination test results are to be reported in whole number

- Labelling of seed packages under seed Act is compulsory
- Test weight is decided based on the seed weight of 100 seeds
- Boerner divider work on the principle of Gravitational force
- In purity analysis wings is considered as inert matter
- Samaras without true seed in purity analysis is known as Pure seed
- The standard method for moisture estimation is Oven dry method
- Blotter test is recommended by ISTA for analyzing seed health
- The pure seed fraction recommended for groundnut is 96 %
- The pure seed fraction recommended for mustard is 97 %
- The pure seed fraction recommended for maize inbred /composite/ synthetic is 96 %
- The pure seed fraction recommended for maize single cross and hybrids are 98%
- The pure seed fraction recommended for cereals is 98%
- Father of Seed Testing Frederick Nobbe
- The first terminology used by Nobbe to explain seed vigor is Shooting strength
- The earliest vigor test is Brick gravel
- The light intensity which favours the seed germination is 660 to 700 nm
- The wavelength of 280nm and below is lethal to plants
- Number of notified Seed Testing Laboratory at Tamil Nadu is 8
- All gymnosperms show Epigeal germination
- During germination, rate of imbibitions increases with increased temperature
- When compared to laboratory germination test, rate of imbibitions in soil is high or equal
- Imbibition is a Physical process
- Phenol test is conducted in wheat
- KOH test is conducted to identify the varieties
- In Seed Testing Laboratory dividers are used to obtain working sample

- Boerner divider is working on the principle of gravitational force
- The compound associated with Tetrazolium test is Formazon
- The enzyme associated with Tetrazolium test is dehydrogenases
- Cold test is considered as best method of estimating vigour in Corn
- Cool germination test is a vigour test for corn
- The type of germination in onion is Epigeal
- Decorticator is used for shelling pods in Groundnut
- Debearders are used in processing of Oats
- Blending is permitted in case of same varieties
- The temperature limit for heated air drying is 1100 F
- First phase of processing is conditioning and cleaning
- Conditioning and pre cleaning improves physical purity
- In processing the role of scalper is rough cleaning
- Scalper favors the cleaning the seeds which contain high trash
- Indent cylinder separator, separates the seeds based on length
- Basic grading is separation of seeds based on size
- Specific gravity separator separates seeds based on seeds differing in density
- Seed differing in size can be separated using seed grader
- Hard seed coat can be smoothened by using scarifier
- Round and flat seeds in a seed lot can be separated by draper belt
- Dodder mill separates seeds based on surface texture
- Seeds which differ in color can be separated by electronic color shorter
- Magnetic separator separates seeds based on the affinity towards liquid
- Excessive glumes in forage seeds can be removed by debearder
- Electrostatic separator, separates seeds based on their difference in electrical conductivity
- Clusters of seeds can be separated by Huller
- Troublesome seed appendage scan be removed by hammer mills
- Barley awns are removed by using de bearder

- In air screen cleaner fast shaking of sieves will help in removal of cleaning chaffy seeds
- Low shake speed in a air screen cleaner used to process round shape seeds
- Seeds differing in length can be separated by Indent cylinder separator
- In an air screen cleaner top screen does the function of scalping
- In an air screen cleaner middle screen does the function of grading
- The first step of seed processing is basic cleaning
- Normally seed extraction is not applicable to diptero carpus
- The ideal location for placing scalper in a processing complex is entry room
- From pinuscones seeds are removed using thumbler
- During processing the genetic purity assessment is taken in maize
- During processing care should be taken to avoid mechanical damage
- Empty seeds of Emblica officinalis can be removed by water floatation
- The seeds are not extracted from fruits during processing in teak
- Color grading could be adapted to separate good seeds in Acacia mellifera
- Wet extraction is the common practice of extraction in bittergourd
- The temperature limit for heated air drying is 1100 F
- Clusters of seeds can be separated by Huller
- Cuscuta seeds are removed from Lucerne using magnetic separator
- Iron powder is used in separation of good seeds in peas
- Mirror is used in separation of good seeds in electronic color sortor
- Round and flat seeds of peas can be separated by using spiral separator
- The book on "Seed Processing" was written by Gregg
- Storage life of seed is influenced by storage environment, genetic make up of seed and moisture content of seed
- Darkening of seed coat during storage is due to Oxidation of Phenol
- The desiccant used for storage of seed is Silica gel

- The chemical used for Mid-storage correction is KH2PO4
- Bruchid is a Field carry over storage pest
- Liquid Nitrogen is used in Cryopreservation
- The safe moisture content for storage of paddy seeds under ambient condition is 13%
- Coimbatore is classified as fair storage place
- Best storage place in India is Mumbai
- Nomo graph is useful to predict the Seed viability period
- Deterioration is more in non-dormant seed
- Electrical Conductivity of seeds steep water indicates Membrane integrity
- On proper seed storage seed quality is maintained
- Size of base collection of heterogeneous material to store in a gene bank is 12,000 seeds
- Size of base collection of homogenous material to store in the gene bank is 4000 seeds
- Thumb rule for safe seed storage has been developed by Harrington
- Liquid nitrogen is used for Cryogenic storage
- Dr. Basu, the Seed technologist is specialized In Mid-storage correction
- The chemical used for seed hardening is KCl
- Storage life of seed is influenced by Temperature & Relative Humidity
- The book on seed viability was written by E. H. Roberts
- Hardened seed performed well in rainfed sowing
- Priming is much suited pre sowing seed treatment for sub optimal condition
- Stratification is the treatment given to break the immature embryo dormancy
- Physiological dormancy is due to absence of PGR
- Dormancy of Apple seeds is broken by cold stratification
- Seed pelleting is Pre-sowing seed treatment

- Infusion of Plant Growth Regulator into the seed through organic solvent dry permeation
- Pancha gavya is a Bio product used for multipurpose
- During dry permeation the seed moisture content is no change
- Seed treatment chemical against loose smut of wheat is Carboxin
- The priming agent, which does not penetrate the cell wall, is PEG
- The priming agent which penetrate the cell wall is mannitol
- Hallow heart in pea occurs due to Molybdenum deficiency
- Presence of pathogen on the surface of the seed is termed as Contamination
- Presence of pathogen inside the seed is termed as Infestation
- Presence of sclerotia spores are termed as Admixture
- Concomitant contaminant of seed is infestation
- Externally seed borne pathogen is Cercospora kikuchii
- In Sorghum. Mycelium of Fusarium monliforme is present in inner seed coat
- The causal organism of downy mildew in maize is present in Upper part of the plumule
- Brown spot in rice is transmitted through Seed
- Fusarium is a field fungi and storage fungi
- Lesser grain borer is the storage pest of Rice
- Musty odour and dust in seed storage godown indicates presence of Khapra beetle
- The storage pest of wheat is Lesser grain borer
- Objectionable disease in sesame is Downey mildew
- Scale is the objectionable insect pest of Tapioca
- PCNB media is most useful in detection of fungal species Penicillium
- Bacterial pathogen associated with seed is detected through ELISA
- Kernel bunt of wheat can be identified by NaOH soaking test
- Double diffusion test is used for the deduction of seed borne nematodes
- Black color discoloration in wheat seed is due to the infection of Tilletia indica

- Collateral host of pearl millet ergot is Cenchrus
- Ephestia cautella in brinjal is a Secondary pest
- Ascochyta blight of bean is caused by Ascochyta phasolorun
- Cautella larvae is attacked by Braconhebetor
- Scientific name of lesser grain borer Rhizopertha dominica
- Rhizopertha dominica is an internal feeder
- In rice seeds curvularia causes eye shaped discoloration
- Exerohilium oryzae in rice causes Black discoloration
- Light pink discoloration in seeds is due to Fusarium monliforme
- Presence of dark brown spot in seed is due to Alternaria alternate
- Fruit rot of chilli is caused by Colletotrichum Capsici
- Fruit rot disease of chilli is Seed borne
- Brine solution is recommended against seed infection
- As per schedule of plant quarantine regulation 1981, the quarantine pest of mango is Seed weevil
- Leakage of organic metabolites from seed induces the Growth of microorganisms
- Death of a plant part due to pest attack of Necrosis
- Wheat variety which is susceptible to loose smut is Sonalika
- Theviabilityperiodofpeapodspotpathogenis6years
- Black rot of cabbage is caused by bacteria
- A sudden broke down of bacterial blight of rice was occurred from Bihar during 1962
- Seedling blight of maize was first observed in India during 1957
- Allium seeds are given with Phytosanitary certificate for the disease smut
- Uricystis cepulae causes onion smut
- Without Phytosanitary certificate for smut disease the onion seeds cannot be exported to Pakistan
- Presence of pathogen on the surface of the seed is termed as Contamination
- Presence of pathogen inside the seed is termed as Infestation

Memory based Previous ARS Questions

1. Describe different types of male sterility in plants. How cytoplasmic genetic male sterility has been utilized/ gives suitable examples.
2. Explain the various hypothesis of heterosis. Which of the hypothesis is more widely accepted and why?
3. Compare the efficiency of different schemes for population improvement in cross pollinated crops. Which scheme do you prefer and why?
4. Describe the types of chromosomal variations in crop plants. How these variations are useful for a crop breeding programme?
5. 5. Why multi location testing is necessary in varietal improvement programme?
6. Discuss the techniques and methods for increasing accuracy of field experiment in plant breeding?
7. In relation to crop improvement discuss the importance of conservation of biodiversity and also support how In-situ and Ex-situ conservation of genetic resources are useful for crop improvement.
8. Give significance of host plant resistance to insects. Outline breeding methodology for host plant resistance
9. Describe the usefulness of biotechnology in agriculture. Give specific of crops where this technique has been used to introduce traits of economic importance.
10. State with specific examples the accomplishments made through plant breeding in India in the last 50 years for different objective. Also comment on the undesirable consequences.
11. What are the components of genetic variation? Outline a biometrical method for determining gene action. How this information is utilized in crop improvement.
12. What do you understand by the term heritability/ describe the different methods to estimate heritability and comment on its utility to the plant breeder.

13. Mention the role of plant biotechnology in crop improvement and explain with examples the utility of these techniques to achieve specific objectives.
14. What is a QTL? Describe briefly the role of molecular markers in crop improvement.
15. Define the heterosis and explain its genetic basis. Mention in detail the exploitation of heterosis in rice and sunflower.
16. Define recurrent selection. Give a brief account of various types of recurrent selection employed for specific situations.
17. Name different intra and inter-population improvement methods. Describe full sib reciprocal recurrent selection in brief with the help of diagram.
18. Describe backcross method of breeding assuming that the character under transfer is governed by single recessive gene.
19. Comment on the statement, "Plant breeding will continue to play a crucial role in crop improvement in 21st century".
20. Explain the fundamental principles on which single seed descent method is based. What are its merits and demerits?
21. Enlist the various systems of mating. Explain in brief the genetic consequences of these systems.
22. Outline the procedure of seed production of double cross hybrid maize utilizing CMS system
23. The yield of a synthetic/ composite variety should not decline provided the population is large enough and mating is at random/ comment on the statement and discuss the principle behind it.
24. What is marker assisted selection? Discuss different kinds of molecular markers with their advantages and disadvantages.
25. Discuss structural chromosomal aberrations
26. Discuss the genetic and molecular basis of heterosis
27. Define the term 'stability'. What is its mechanism?
28. Explain briefly the various measures of stability of genotype used in plant breeding
29. Give the molecular approach to breeding for wider adaptability.
30. Discuss the genetic basis of vertical and horizontal resistance

31. What is seed certification? Why is it essential to have seed certification programme?
32. What are the centers of origin? How are primary, secondary and tertiary centers of origin determined?
33. List the various centers of origin of crop plants giving the examples for each centers, as given by Vavilov
34. What is recombinant DNA molecule? How is the polymerase chain reaction (PCR) such a powerful tool for use in DNA analysis?
35. What is the significance of recombinant DNA and gene cloning technologies to plant breeders?
36. How are the Restriction Endonucleases used to construct rDNA molecules in vitro? Explain with the help of diagram only.
37. What is abiotic stress tolerance in crops? What is its significance in agriculture? Discuss the development of abiotic stress tolerant genotypes through an integrated participatory molecular breeding approach
38. Discuss the multi- disciplinary approach of plant breeding. Plant breeding is now modern science. Prove it with suitable examples. Enumerate the land mark achievements of it, detailing its scope for crop improvement.
39. Give the parameters to identify Apomictic plants. Describe the genetics of apomixes. Also discuss the scope of Apomixis for gainful utilization in breeding of field crops.
40. Enlist the various centers of origin of cultivated plants citing few important examples of them, which originated there. Briefly describe the concept of centers of origin (primary and secondary). How is this concept helpful in plant breeding?
41. Define gene pool. Enlist the various procedures involved in germplasm collection. Also discuss the strategies of germplasm collection and conservative programme.
42. Discuss the various types of male sterility used in crop improvement. Also discuss the scope of two line system for hybrid seed production. Briefly describe the achievements citing some examples.
43. How inbred lines are produced in self in compatibility system? How two locus system of incompatibility are helpful in producing hybrids? Describe the problems associated with self incompatibility in hybrid seed production.

44. Briefly describe the multiple factor hypothesis of polygenic inheritance. How did the findings off Nilson- Ehle show that the quantitative characters are inherited according to Mendelian inheritance?

45. What are the various components of genetic variance/ how would you estimate the values of D and H components from the variance of different generations.

46. What is the effect of linkage relationship of genes at different loci on possible gene recombination that obtained through hybridization? How different genetic phenomena influence the planned hybridization?

47. Discuss the significance of induced mutations in plant breeding. Give the advantages of tissue culture mutagenesis. What are the main limitations of using mutation breeding?

48. Enlist the causes of somaclonal variations. How somaclonal variations are induced and selections made? Discuss the significance of somaclonal variations in crop improvement.

49. How haploids are produced in in vitro? Compare the double haploids with pure lines produced through conventional breeding methods. Why double haploids are not much helpful for production of inbred line in cross pollinated species?

50. What are the different intra- population improvement schemes? Discuss their comparative efficiencies in improving the population. What role these schemes have played in improvement of the cross pollinated crops?

51. Discuss the methods of breeding for disease resistance. Describe the backcross method of breeding if the disease is governed by a recessive gene.

52. List the basic difference between prokaryotic and eukaryotic gene control.

53. What are the significant uses of genome analysis in crop plants?

54. Differentiate between missense, nonsense, samesense and frameshift mutations.

55. Explain the cis-trans test. What is its utility and limitations?

56. Defend or refute the following statement citing suitable examples. Transposable elements are junk DNA with no useful biological function.'

57. Define maternal inheritance. How does it differ from cytoplasmic inheritance?
58. How are deletions used for gene mapping?
59. Briefly describe the features of polytene chromosomes.
60. What is Mendel's law of independent assortment? Explain why Mendel did not encounter the problem of linkage in his experiments.
61. Describe sex determination systems in plants.
62. Describe the experiment that proved that DNA replication is semi conservative.
63. Discuss various features of genetic code.
64. What is the significance of complimentarity and antiparallel nature of the DNA double helix?
65. Mention the cytological and genetic procedures to detect chromosomal interchange in plants.
66. How is mode of pollination determined in a species? Describe the procedure for determination the extent of cross pollination in a species.
67. Discuss the patterns of evolution in cultivated crops, giving suitable examples.
68. List different types of Apomixis. What are the indicators of apomixes?
69. What are the prerequisites of backcross breeding? Illustrates diagrammatically the procedure to transfer a recessive gene from a donor to the recurrent parent.
70. Describe the procedure for production of alien substitution lines, citing suitable examples.
71. Distinguish between macro mutation and micro mutation. Briefly describe the procedure to develop high yielding cultivars using micro mutation breeding approach.
72. Compare the effectiveness of various recurrent selection schemes for population improvement.
73. Mention the barriers to wide hybridization and the scientific methods to overcome such barriers.
74. Briefly describe the basic features of hardy- Weinberg equilibrium' and mention its applications in genetics and plant breeding.

75. Compare the strengths and limitations of F2:3 and RIL mapping populations in relation to QTL mapping in crop plants.
76. Define a biparental cross. Give different designs used for biparental mating and their relative merits and demerits.
77. Write 'multilines' and briefly describe how these lines are produced. Why are multilines not so popular?
78. Define briefly the constraints in breeding for drought stress tolerance in crop plants.
79. Discuss briefly the various industrial and specialty uses of maize.
80. List various species of minor millets. Highlight the nutritional importance of minor millets, particularly for eradication of 'hidden hunger'.
81. What is the center of origin of sugarcane? List various species of Saccharum and comment on their utility in sugarcane breeding
82. What are the different cultivated species of rice? Mention the salient characteristics and commercial importance of basmati rice.
83. Trace the evolutionary history of bread wheat.
84. Discuss the origin and various uses of cultivate soybean.
85. Give the botanical classification and morphological description of pigeon pea.
86. Illustrate the U triangle and discuss its significance.
87. Explain with the suitable examples, the roles of mutation, polyploidy and chromosomal aberrations in evolution of crops.
88. Describe the procedure involved in producing synthetic and composite varieties.
89. What are the components of genetic variance? How they are estimated? Explain how these values are used in breeding procedures?
90. Compare breeder seed and certified seed and explain how seed certification is carried out.
91. Differentiate domestication and pre breeding
92. How are chloroplast and mitochondrial genome different from nuclear genome?
93. Describe complementary and supplementary gene interactions with suitable examples.

94. Explain the mechanism through which 5-bromouracill, nitrous acid and EMS bring about mutations.
95. Diagrammatically show the differences and consequences of paracentric inversion and pericentric inversion.
96. Briefly describe post transcriptional processing in eukaryotes. Explain
97. How are monosomics used for gene mapping?
98. Explain how the present day maize and sorghum were derived from their progenitors.
99. Explain briefly the procedures used for stability analysis
100. What are the applications of mitochondrial genome analysis for crop improvement?
101. How do you define 'adaptability' and 'stability' in relation to breeding?
102. Describe the salient features of lampbrush chromosomes (with diagrams), and discuss their utility in genetic studied.
103. What is meant by comparative mapping? Mention its utility.
104. What are the various types of repetitive DNA in genome?
105. Differentiate between heterochromatin and euchromatin.
106. List out the various types of chromosome banding techniques and explain any one.
107. What are 'multiple alleles'? Explain the concept with the suitable examples.
108. What is 'Linkage Disequilibrium'? Briefly describe the various factors influencing linkage disequilibrium.
109. What are the applications of genetic diversity analysis in plant breeding?
110. Mention the advantages of molecular markers offer in relation to diversity analysis?
111. What is meant by marker assisted selection? Explain how DNA based markers can help in pyramiding genes for disease resistance, using suitable examples.
112. Explain the procedure of Biosafety evaluation and release of transgenic cultivar in the Indian context.

113. How do you define an 'operon'? Explain briefly the inducible and repressible operons, citing suitable examples.
114. Mention different types of DNA based markers and their characteristics?
115. Differentiate between 'plant breeder's right', breeder's exemption and farmers right under the PPV and FR act of India.
116. What are the salient reproductive systems observed in crop plants? What mechanisms/features are responsible for cross pollination? Which system pollination do you think is the most evolved? Justify.
117. Genetic differences between prokaryote and eukaryote
118. Differences between correlation and regression coefficient
119. Why heterosis is less in self pollinated crops rather than cross pollinated crops
120. Why correlation coefficient between traits varies with respect to different research studies
121. Differences between pleiotrophy and polygeny
122. SNP applications in plant breeding
123. High iron rice
124. Hybrid development in QPM
125. Different types of male sterility
126. MARS
127. How can polyunsaturated fatty acid can be increased in linseed
128. NBPGR, ICRISAT and IIRR
129. Population improvement approaches
130. Breeding value
131. Diallel mating type
132. NCD differences
133. Eberhart and Russell model for stability analysis
134. VR vs HR
135. Multiline approach
136. Structural and numerical aberrations
137. Irradiation-translocation

138. Genetic stocks in barley to study translocations
139. Heterosis in F_2 and mean performance of F_2
140. Transgenics-impact on environment
141. Transgenics in different crops for different traits
142. Different traits for drought tolerance
143. Differences between inbred vs clone vs pureline
144. Drought avoidance vs drought tolerance
145. Double haploids its utilization
146. Different theories of heterosis
147. Different genetic variances
148. How SSD and bulk breeding methods helpful to breeder
149. Gene regulation in prokaryotes and eukaryotes
150. Complementary epistasis
151. Number of genotypes with 4 heterozygous loci and 2 homozygous loci
152. Chromosome substitution using aneuploid types
153. Clonal selection in sugarcane breeding
154. What is ideotype concept and explain the principles of ideotype
155. What is the role of breeder in seed production chain
156. Different methods of recurrent selection and explain reciprocal recurrent selection
157. FISH
158. Different methods of evolution
159. Strategy to reduce resistance of bollworms for Bt cotton
160. Which is the most commonly useful design for breeders. What is the use of F-test.
161. Explain about the ABD and how it is useful for breeders
162. How meristem culture is different with other tissue culture methods
163. Crossing pattern of different half diallel mating design
164. Is there any difference between biodiversity act and plant variety protection

165. What are the advantages in mendel research experiments
166. MAB
167. Understanding about gene from Mendelian to molecular approach

WRITE SHORT NOTES ON:

- Stability parameters
- Stadlers gamete selection
- Haploids
- What are Okazaki fragments?
- Explain the significance of degeneracy and wobble hypothesis.
- DNA is the genetic material, not proteins
- One gene- one enzyme hypothesis
- B chromosomes
- Breeding applications of interchanges
- Law of homologous series of variation
- Environmental sensitive genetic male sterility
- Evolution of allopolyploids
- Participatory plant breeding
- Single seed descent and bulk method
- Two line and three line methods of hybrid production
- Synthetic variety and synthetic seed
- Hybrid ad vybrid
- Vertical and horizontal resistance
- UPOV
- IPGRI
- QPM (Quality Protein Maize)
- Bt. Cotton
- Golden rice
- Totipotency
- Isogenic lines
- Alkylating agents

- Land varieties
- Somatic embryogenesis
- Vertifolia effect
- Differential hosts
- Multilines
- Gene for gene relationships
- Plant breeders' right
- Patent
- Test of significance
- Phenotypic and genotypic correlations
- Probability
- Geographical indicators
- Material transfer agreement for germplasm
- Genetic transformation in plants
- Cryopreservation bank
- Vertical and horizontal resistance
- Regression
- F-test
- Incomplete block design
- Line x tester analysis
- Apomixes
- Recurrent selection
- Augmented design
- Plant breeders right and variety protection
- QTLs
- RFLP
- Microsatellite markers
- Metroglyph analysis
- Over dominance hypothesis
- G x E interaction
- SSD

- Bt cotton
- Transgenic crops with virus resistance
- QPM
- 0-0 mustard
- Golden rice
- Biofortification
- cDNA libraries
- Methods of DNA sequencing

9 789359 191485

Printed at Repro India Ltd.